Geoplanet: Earth and Planetary Sciences

For further volumes:
http://www.springer.com/series/8821

Karol Kuliński · Janusz Pempkowiak

Carbon Cycling in the Baltic Sea

Springer

Karol Kuliński
Institute of Oceanology
Polish Academy of Sciences
Powstancow Warszawy 55
81-712 Sopot
Poland
e-mail: kroll@iopan.gda.pl

Janusz Pempkowiak
Institute of Oceanology
Polish Academy of Sciences
Powstancow Warszawy 55
81-712 Sopot
Poland
e-mail: pempa@iopan.gda.pl

The GeoPlanet: Earth and Planetary Sciences Book Series is in part a continuation of Monographic Volumes of Publications of the Institute of Geophysics, Polish Academy of Sciences, the journal published since 1962 (http://pub.igf.edu.pl/index.php).

ISSN 2190-5193
ISBN 978-3-642-19387-3
DOI 10.1007/978-3-642-19388-0
Springer Heidelberg Dordrecht London New York

e-ISSN 2190-5207
e-ISBN 978-3-642-19388-0

Library of Congress Control Number: 2011940783

Cover design: deblik, Berlin

Printed on acid-free paper

Springer is part of Springer Science+Business Media (www.springer.com)

Series Editors

Managing Editor

Advisory Board

Preface

'Global change' is a frequently used, collective description of a number of processes and symptoms, largely caused by anthropogenic activity, that affect the Earth's ecosystems and endanger mankind. Global change is likely to exert changes in the environment to an extent that has not been witnessed historically.

Climate Warming is one of the global change processes that are certain to strongly affect the environment in the nearest future. There are several factors, often referred to as forcings, that cause the increase of the Earth's atmosphere average temperature. The most important one, responsible for some 60% of the increase, is the growing concentration of carbon dioxide in the atmosphere. This is, usually, attributed to the so-called anthropogenic emissions of carbon dioxide (CO_2) to the atmosphere. The emissions are a direct result of fossil fuels burning, which is due to increasing demand for energy, necessary to maintain or increase civilization standards of mankind. Thus, it is difficult to imagine that, in the nearest future, the atmospheric concentration increase rate of CO_2 might change. Since there is a direct link between the Climate Warming and CO_2 concentration in the atmosphere, all model studies, aimed at assessing the extent of the warming, need to be based on future concentrations of CO_2 in the atmosphere. That is why the quantification of factors determining carbon dioxide concentration in the atmosphere is of interest to and a challenge for contemporary science.

Therefore, investigations are necessary to quantify, as accurately as possible, CO_2 sinks and sources to the atmosphere.

The global ocean is believed to absorb some 40% of the anthropogenic CO_2 emissions. This is an effect of the CO_2 uptake by both the open ocean and shelf seas. Without quantification of both sinks it is impossible to improve the ocean CO_2 absorption estimates, and future CO_2 concentration changes. The quantification requires a major effort on the part of researchers engaged in investigating carbon cycling in the marine environment.

This book provides results of the research on CO_2 cycling in the Baltic Sea that has been carried out within the last few years. The Baltic—a land locked sea located in the temperate climate, has been a subject of oceanographic, biological, physical and chemical investigations for a hundred years or so. The results of the studies combined with our own data gathered with the aim of characterizing carbon reservoirs and fluxes in the sea, have been used to define carbon cycling in the Baltic. The cycling, quantified separately for inorganic carbon and organic carbon, is influenced primarily by carbon discharged to the sea with river run-off, exchange with the North Sea, deposition to- and return flux from- the bottom sediments, and exchange of carbon with the atmosphere. The last component determines the sea trophic status. The outcome of the studies classifies the Baltic as an area of slightly heterotrophic (degassing CO_2) to neutral (neither degassing nor absorbing CO_2) status.

The authors believe that the book contents, apart from clarifying the Baltic carbon trophic status, will improve understanding of the role shelf seas play in ocean carbon cycling, and the mechanisms of boundary processes responsible for carbon budgets in the shelf seas.

Contents

1 Introduction 1
References 3

2 Climate and Carbon Cycle 5
2.1 Climate: Definition and Prerequisites 5
2.1.1 Natural Factors Controlling the World's Climate 6
2.1.2 Climate Change and Anthropogenic Influences 8
2.1.3 Global Consequences of the Climate Change 13
2.1.4 Climate Change Consequences in the Baltic Sea Region 17
2.2 Global Carbon Cycle 19
2.2.1 Carbon Reservoirs 19
2.2.2 Scheme of the Global Carbon Cycle in the Environment 20
2.2.3 Carbon Cycle in the Marine Environment 22
2.2.4 Role of Shelf Seas in the Atmospheric Carbon Dioxide Absorption 26
2.3 Carbon Cycling in the Baltic Sea 28
2.3.1 Baltic Sea Characteristics 28
2.3.2 Carbon Reservoirs in the Baltic Sea 30
2.3.3 Carbon Sources and Losses in the Baltic Sea 30
References 37

3 Research on Carbon Cycling in the Baltic: Aims and Scope 49
3.1 Aims of the Studies 49
3.2 The Concept of the Study: Quantitative Assessment of the Baltic Sea Carbon Cycle 49
3.2.1 The "Box Model" 49
3.2.2 Significant Carbon Fluxes in the Baltic Sea 51

3.3 The Scope of Studies 52
References 53

4 Research on Carbon Cycling in the Baltic: Quantification of the Carbon Fluxes 55
4.1 Carbon Exchange Between the Baltic Sea and the North Sea . . . 55
4.1.1 Dissolved Inorganic Carbon 58
4.1.2 Dissolved Organic Carbon 61
4.2 Carbon Inflows with River Runoff. 64
4.3 Carbon Deposition to the Baltic Sea Sediments. 66
4.3.1 Carbon Return Flux from the Bottom Sediments. 67
4.3.2 Labile Organic Matter Decomposition in Sediments 71
4.4 Carbon Inflow to the Baltic with Precipitation 76
References 78

5 Research on Carbon Cycling in the Baltic: Discussion 81
5.1 The Baltic Sea Carbon Budget 81
5.2 Factors Determining Carbon Budget in the Baltic Sea 86
5.3 The Budget Uncertainty 90
5.4 Deposition and Origin of Organic Matter to the Baltic Sediments. The Case of Cores Dated with ^{210}Pb and ^{137}Cs 93
5.5 Influence of the Climate Change on the Carbon Resources and Fluxes in the Baltic Sea 102
References 104

Appendix 109

Index 125

List of Abbreviations

BACC	The book "Assessment of Climate Change for the Baltic Sea Basin"
Baltic-C	BONUS funded project
BOD	Biological Oxygen Demand
BONUS	Funding agency for Baltic Sea research
BSH	the Bundesamt für Seeschifffahrt und Hydrographie
BSH CMOD	Hydrostatic 3D circulation model operated by DMI
CFCs	Chlorofluorocarbons
C_{org}	Organic carbon
DIC	Dissolved inorganic carbon
DMI	Danish Meteorological Institute
DOC	Dissolved Organic Carbon
ENSO	El Niño-Southern Oscillation
HELCOM	Helsinki Commission, Baltic Marine Environment Protection Commission
HIRLAM	High Resolution Limited Area Model
IC	Inorganic carbon
ICES	International Council for the Exploration of the Sea
IGBP	International Geosphere-Biosphere Programme
IPCC	Intergovernmental Panel on Climate Change
IR	Infrared Radiation
JGOFS	Joint Global Ocean Flux Study
K_H	Henry's law constant
LSA	Linear Sediment Accumulation
MAR	Mass Sediment Accumulation
MB	End member station located in the Arkona Basin (Baltic Sea)
NAO	North Atlantic Oscillation
NDIR	Non-Dispersive Infrared
OC	Organic Carbon
pH	A negative decimal logarithm of the hydrogen ion activity in solution

PIC	Particulate Inorganic Carbon
POC	Particulate Organic Carbon
POM	Particulate Organic Matter
R	Correlation coefficient
R^2	Determination coefficient
RSD	Relative Standard Deviation
S	Salinity
SD	Standard Deviation
TIC	Total Inorganic Carbon
TOC	Total Organic Carbon
UV	UltraViolet

Chapter 1
Introduction

Global carbon cycle plays a key role in the matter and energy fluxes of the Earth system. Carbon cycle links all the components of the Earth system: land, ocean and the atmosphere (Emerson and Hedges 2008). Since respiration and photosynthesis are the two fundamental processes in the carbon cycle, CO_2 becomes the crucial carrier of the matter in the environment.

Last few decades of mankind history are characterized by a rapid socio-economic transformation and, resulting from this, global changes. The greenhouse effect is one of the most appreciable and best recognized among all the global changes symptoms. It is believed that the mankind emits to the atmosphere approximately 22 Pg of carbon dioxide (CO_2) each year. The anthropogenic CO_2 emission causes alterations of carbon fluxes between the Earth's compartments and thus disturbs natural carbon cycling (IPCC 2007). Approximately 40% of the anthropogenic CO_2 is stored in the ocean. Thus, understanding the role of the marine carbon cycle in global and regional climate regulation is one of the major objectives of present-day oceanography.

The so-called biological pump plays a key role among the processes that govern the uptake of atmospheric CO_2 by the ocean. Biological pump is forced by the photosynthesis that transforms CO_2 into organic constituents of phytoplankton cells. After death and lyses of the cells, organic components are spilled out to become dissolved and particulate components of seawater. Significant part of the dead organic matter is mineralized in the water column, while its portion is deposited into sediments where is further mineralized. A fraction of the deposited organic matter escapes mineralization and is eventually buried into subsurface sediments. It has been estimated that, globally, from 0.01 to 1% of organic carbon originating from the primary production is buried in marine sediments and thus is excluded from the immediate carbon cycle (Rullkötter 2006).

Shelf seas play a key role in the global oceanic fluxes of matter and energy (Thomas et al. 2009). Although they make up a little over 7% of the global sea surface and less than 0.5% of the ocean seawater volume, shelf seas are

K. Kuliński and J. Pempkowiak, *Carbon Cycling in the Baltic Sea*,
Geoplanet: Earth and Planetary Sciences, DOI: 10.1007/978-3-642-19388-0_1,

responsible for 15–30% of marine primary production and, as much as, 80% of organic matter burial (Walsh 1991; Borges 2005; Bozec et al. 2005; Chen and Borges 2009). These features of shelf seas are caused by the high biological activity they support, activity that is driven by nutrients inputs from all of the adjacent environments (Gattuso et al. 1998; Pätsch and Kühn 2008; Thomas et al. 2009).

As a consequence of the high biological productivity, most global shelf seas are believed to act as net sinks for anthropogenic CO_2 (e.g., Chen et al. 2003; Borges et al. 2005; Chen and Borges 2009). Moreover, the CO_2 loads absorbed by shelf seas exceed those reported for the open ocean (Chen and Borges 2009; Takahashi et al. 2009). On the other hand it has been suggested recently that, in contrast to open shelf seas, some near-shore zones are identified as sources of CO_2 to the atmosphere (Chen and Borges 2009; Liu et al. 2010a). This is due to the large organic matter loads from land undergoing mineralization in well oxygenated seawater. Consequently, detailed studies of the carbon cycle in shelf seas are still required in order to clarify their role in the global carbon cycle. Although several attempts have been made to quantify the role of shelf seas in global CO_2 fluxes (Andersson and Mackenzie 2004; Thomas et al. 2004; Tsunogai et al. 1999), validation of the conclusions from these studies must be based on compilations of the results of detailed local studies. These enable the multifarious locally specific processes influencing CO_2 exchange between seawater and the atmosphere to be taken into consideration (Borges 2005; Borges et al. 2005; Chen and Borges 2009).

The Baltic Sea is a spatially and temporally highly diverse ecosystem (Dippner et al. 2008; Helcom 2009). This is the reason behind the significant discrepancies in the CO_2 air-sea exchange results reported in the literature (Ohlson 1990; Thomas and Schneider 1999; Thomas et al. 2003; Algesten et al. 2004, 2006; Kuss et al. 2006; Wesslander et al. 2010). Although the Bothnian Sea and the Gulf of Bothnia are considered to be a net source of CO_2 to the atmosphere (Algesten et al. 2004, 2006), the Baltic Proper with adjacent gulfs is believed to be an area within the European shelf, where the atmospheric CO_2 is absorbed despite the proven high spatial and temporal variability of CO_2 partial pressure in seawater (pCO_2) (Ohlson 1990; Thomas and Schneider 1999; Thomas et al. 2003; Kuss et al. 2006; Chen and Borges 2009). The calculated, final direction of the CO_2 exchange through the sea surface seems to be related to the boundary processes, specifically sinks and sources of carbon in the sea (Thomas et al. 2003). Although the biogeochemical processes governing the carbon cycle in the Baltic Sea are well defined qualitatively, quantification of carbon fluxes still requires detailed investigations.

The results regarding pCO_2 measurements in the Baltic, reported in the literature, were collected mostly at stations located in the open waters of the Baltic Sea. The near-shore zones and areas adjacent to river mouths have been largely omitted from pCO_2 measurements. However, these regions of the Baltic Sea are likely to be of special importance for CO_2 cycling, since it has been demonstrated worldwide that near-shore zones and river mouths are important sources of CO_2 to the atmosphere due to the high loads of terrestrial carbon (Frankignoulle et al.

1998; Borges 2005; Chen and Borges 2009; Liu et al. 2010b). The rivers flowing into the Baltic Sea drain an area that is more than four times larger than that of the sea itself. Moreover, the water volume the rivers discharge annually to the Baltic Sea amounts to almost 2% of the total Baltic capacity (Lass and Matthäus 2008), and is characterized by, both dissolved and particulate, organic matter concentrations that are several times larger than those in seawater.

The aspects mentioned above were the motivation for the authors to write this book and present a state-of-the-art comprehensive description of all the boundary carbon fluxes that provide the carbon budget for the entire Baltic Sea. The resulting budget indicate that the Baltic, as a whole, acts as a source of carbon dioxide to the atmosphere. However, if the Gulf of Bothnia is excluded, the remaining area turns into a net, strong, absorber of CO_2.

References

Algesten G, Wikner J, Sobek S, Tranvik LJ, Jansson M (2004) Seasonal variation of CO_2 saturation in the Gulf of Bothnia: indications of marine net heterotrophy. Glob Biogeochem Cycles 18, GB4021

Algesten G, Brydsten L, Jonsson P, Kortelainen P, Löfgren S, Rahm L, Räike A, Sobek S, Tranvik L, Wikner J, Jansson M (2006) Organic carbon budget for the Gulf of Bothnia. J Mar Syst 63:155–161

Andersson AJ, Mackenzie FT (2004) Shallow-water oceans: a source or sink of atmospheric CO_2? Frontiers Ecol Environ 2:348–353

Borges AV (2005) Do we have enough pieces of the jigsaw to integrate CO_2 fluxes in the coastal ocean? Estuaries 28:3–27

Borges AV, Delille B, Frankignoulle M (2005) Budgeting sinks and sources of CO_2 in the coastal ocean: diversity of ecosystems counts. Geophys Res Lett 32:L14601

Bozec Y, Thomas H, Elkalay K, de Baar HJW (2005) The continental shelf pump for CO_2 in the North Sea-evidence from summer observation. Mar Chem 93:131–147

Chen C-TA, Borges AV (2009) Reconciling opposing views on carbon cycling in the coastal ocean: continental shelves as sinks and near-shore ecosystems as sources of atmospheric CO_2. Deep-Sea Research II 56:578–590

Chen C-TA, Liu K–K, Macdonald R (2003) Continental margin exchanges. In: Fasham MJR (ed) Ocean Biogeochemistry. Springer, Berlin, pp 53–97

Dippner JW, Vuorinen I, Daunys D, Flinkman J, Halkka A, Köster FW, Lehikoinen E, MacKenzie BR, Möllmann C, Mohlenberg F, Olenin S, Schiedek D, Skov H, Wasmund N (2008) Climate-related marine ecosystem change. In: The BACC author team (eds) Assessment of climate change for the baltic sea basin. Springer, Berlin, pp 309–377

Emerson SR, Hedges JI (2008) Chemical Oceanography and the marine carbon cycle. Cambridge University Press, Cambridge, p 453

Frankignoulle M, Abril G, Borges A, Bourge I, Canon C, Delille B, Libert E, Théate J-M (1998) Carbon dioxide emission from European estuaries. Science 282:434–436

Gattuso J-P, Frankignoulle M, Wollast R (1998) Carbon and carbonate metabolism in coastal aquatic ecosystems. Annu Rev Ecol Syst 29:405–434

Helcom (2009) Eutrophication in the Baltic Sea. Baltic Sea environment Proceedings 115B, str. 150

IPCC (2007) Climate change 2007: synthesis report. A contribution of working groups I, II and III to the fourth assessment report of the intergovernmental panel on climate change. Cambridge University Press, Cambridge, p 73

Kuss J, Roeder W, Wlost KP, DeGrandpre MD (2006) Time-series of surface water CO_2 and oxygen measurements on a platform in the central Arkona Sea (Baltic Sea): seasonality of uptake and release. Mar Chem 101:220–232

Lass H-U, Matthäus W (2008) General Oceanography of the Baltic Sea. In: Feistel R, Nausch G, Wasmund N (eds) State and evolution of the Baltic Sea, 1952–2005. Wiley, Hoboken, pp 5–43

Liu K-K, Tseng C-M, Wu C-R, Lin I-I (2010a) The South China Sea. In: Liu K-K, Atkinson L, Quiñones RA, Talaue-McManus L (eds) Carbon and nutrient fluxes in continental margins. Springer, Berlin, pp 464–482

Liu K-K, Atkinson L, Quiñones RA, Talaue-McManus L (2010b) Biogeochemistry of continental margins in a global context. In: Liu K-K, Atkinson L, Quiñones RA, Talaue-McManus L (eds) Carbon and nutrient fluxes in continental margins. Springer, Berlin, pp 3–24

Ohlson M, (1990) Some aspects of a budget for total carbonate in the Baltic Sea. In: Proceedings of the 17th Conference of the Baltic Oceanographers, Norrköping, pp 78–83

Pätsch J, Kühn W (2008) Nitrogen and carbon cycling in the North Sea and exchange with the North Atlantic–A model study. Part I. Nitrogen budget and fluxes. Cont Shelf Res 28:767–787

Rullkötter J (2006) Organic matter: the driving force for early diagenesis. In: Schulz HD, Zabel M (eds) Marine Geochemistry. Springer, Berlin, pp 125–206

Takahashi T, Sutherland SC, Wanninkhof R, Sweeney C, Feely RA, Chipman DW, Hales B, Friederich G, Chavez F, Sabine C, Watson A, Bakker DCE, Schuster U, Metzl N, Yoshikawa-Inoue H, Ishii M, Midorikawa T, Nojiri Y, Körtzinger A, Steinhoff T, Hoppema M, Olafsson J, Arnarson TS, Tilbrook B, Johannessen T, Olsen A, Bellerby R, Wong CS, Delille B, Bates NR, de Baar HJW (2009) Climatological mean and decadal change in surface ocean pCO_2, and net sea–air CO_2 flux over the global oceans. Deep Sea Res II 56:554–577

Thomas H, Schneider B (1999) The seasonal cycle of carbon dioxide in Baltic Sea surface waters. J Mar Syst 22:53–67

Thomas H, Pempkowiak J, Wulff F, Nagel K, (2003) Autotrophy, nitrogen accumulation and nitrogen limitation in the Baltic Sea: a paradox or a buffer for eutrophication? Geophy Res Lett 30, GL017937

Thomas H, Bozec Y, Elkalay K, de Baar HJW (2004) Enhanced open Ocean storage of CO_2 from shelf sea pumping. Science 304:1005–1007

Thomas H, Schiettecatte L-S, Suykens K, Koné YJM, Shadwick EH, Prowe AEF, Bozec Y, de Baar HJW, Borges AV (2009) Enhanced ocean carbon storage from anaerobic alkalinity generation in coastal sediments. Biogeosciences 6:267–274

Tsunogai S, Watanabe S, Satao T (1999) Is there a continental shelf pump for the absorption of atmospheric CO_2? Tellus B 5:701–712

Walsh JJ (1991) Importance of continental margins in the marine biogeochemical cycling of carbon and nitrogen. Nature 350:53–55

Wesslander K, Omstedt A, Schneider B (2010) Inter-annual and seasonal variations in the air–sea CO_2 balance in the central Baltic Sea and the Kattegat. Cont Shelf Res 30:1511–1521

Chapter 2
Climate and Carbon Cycle

2.1 Climate: Definition and Prerequisites

Climate, as defined by IPCC (2001, 2007), is the "average weather" at a given period of time and space. In a more statistical understanding, this period of time, ranging in various reports from months to thousands of years, was eventually adopted by the World Meteorological Organization (WMO) to be 30 years.

Temperature, precipitation, and wind are the most commonly used quantities to describe climate and to classify it into specific categories assigned to different parts of the world. There are three basic systems used to categorize weather: (1) Thornthwaite's system, based on the precipitation-to-vapour ratio varying for different areas, (2) system based on air masses identification as the main component forming climate system, (3) Köppen's system (Peel et al. 2007; Martyn 2000; Kalkstein et al. 1996). The latter is most widely applied to describe the world's climate, based on average monthly values of temperature and precipitation; their yearly patterns allowed to differentiate between the five basic climate types:

Type A (tropical)—occurring in the equatorial zone, characterized by continuously high temperatures (monthly averages above 18°C) and abundant, year-round, precipitation;

Type B (dry)—distinctive for vast areas of North Africa, the Arabian Peninsula and Australia, characterized by scarce precipitation and large temperature amplitudes.

Type C (moderate)—characterized by mild winters, occurring in the mid-latitude regions (e.g., Western Europe)

Type D (continental)—mid-latitude zone climate characterized by cool winters (average temperatures of the warmest and the coolest month oscillate above and below 10°C, respectively),

Type E (polar)—typical for high latitude regions, characterized by cool summers (average temperature of the warmest month below 10°C) and cold winters.

K. Kuliński and J. Pempkowiak, *Carbon Cycling in the Baltic Sea*,
Geoplanet: Earth and Planetary Sciences, DOI: 10.1007/978-3-642-19388-0_2,

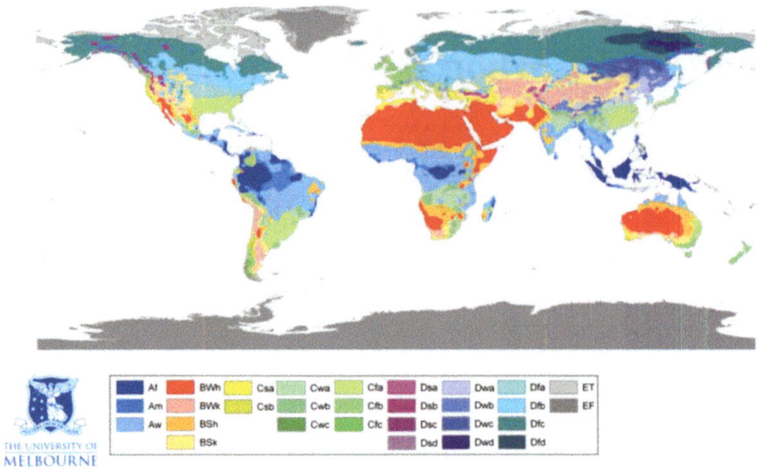

Fig. 2.1 World climate zones distribution (Peel et al. 2007). Climate types: (*a*) tropical, (*b*) dry, (*c*) moderate, (*d*) continental, (*e*) polar; letter abbreviations relate to secondary and tertiary climate classifications

All the types enumerated above are further divided into secondary and tertiary classifications, that, more specifically, distinguish climate on regional and local scales (Fig. 2.1).

2.1.1 Natural Factors Controlling the World's Climate

Specific climate in different parts of the world is an outcome of many factors. Apart from the human influence, that up to recently has been of no or little importance, most of them are natural phenomena (Kożuchowski 1998). Extraterrestrial sources, expressed by the amount of the solar energy reaching the Earth surface and the factors of planetary origin, are directly related to the climate forming processes taking place on the globe (Martyn 2000).

The Earth position in relation to the Sun, facilitating the amount of radiation received from our star, is the basic climate-controlling force. The proximity to the equator, occurrence of daily and yearly radiation cycles generated by the Earth's rotation and orbit, directly influence the Earth's climate (Martyn 2000). Sun–Earth interactions change during longer cycles—the shortest (11 years long) is related to periodically occurring variations in the sunspot number that determine the amount of radiation reaching the Earth surface and its temperature (Rind 2002). Additionally, sunspot cycle seems to be positively correlated with the ocean surface water temperature within the tropical zone (White 2006). Cycles measured in thousands of years are a consequence of the Earth's orbit variability that includes: the orbit precession leading to shifts in the equinox, changes in the tilt of the

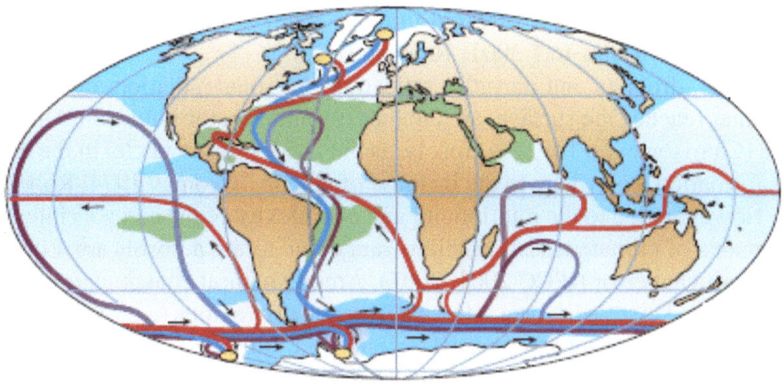

Fig. 2.2 Scheme of the thermohaline circulation (Rahmstorf 2002)

Earth's axis and transformations in the shape of the Earth's orbit, lasting approximately 23, 41, 100 thousand years, respectively.

The differences in a landscape is what determines climate at regional and local scales, varying with height above the sea level, as well as with the distance of the land from the major water bodies that make the transfer between seasons smoother. Both of the above aspects, combined with latitude radiation changes, are causing climate variability that depends on the location (Martyn 2000).

Another important factor controlling climate at the local and regional scales is the relationship between high and low atmospheric pressure systems (Martyn 2000). North Atlantic, for example, is dominated by the Icelandic Low and the Azores High, which decides about the air masses circulation over Europe. Positioning and intensity of both pressure systems change over years, and combined with accompanying events is called North Atlantic Oscillation (NAO; Hurrell 1995). NAO explains about 30 percent of the ocean winter temperatures dynamics north of 20°N (Greatbatch 2000). Similarly, El Niño-Southern Oscillation (ENSO) describes atmospheric circulation over the Pacific Ocean equatorial zone (McPhaden et al. 2006). The range of ENSO repercussions reach vast areas globe wide (Rosenzweig 1994).

Both circulation patterns are directly related to ocean currents pattern (thermohaline circulation)—the World Ocean water circulation process, based on deepwater formation at high latitude regions, mostly in the North Atlantic and around the Antarctica (Fig. 2.2; Rahmstorf 2003). Thus, together with warm surface currents, large amounts of heat are transported to the Arctic region. During the process of gradual water cooling the heat is transferred to the atmosphere (Trenberth and Caron 2001).

The results of a climate model presented by Stocker (2002), that does not involve occurrence of the thermohaline circulation, demonstrated that in this scenario the Southern Hemisphere would cool down insignificantly, whereas the temperature for the Northern Hemisphere, especially in the North Atlantic region, would decrease by several degrees Centigrade. Disruptions of the circulation could have been a possible cause of rapid climate changes during the last glaciations (Ganopolski and Rahmstorf 2001; Rahmstorf 2002).

Aerosols—droplets and particles reflecting the solar radiation and limiting its amount that reaches the surface, are another essential factor influencing the Earth's climate (Charlson et al. 1992). Aerosols, as the condensation nuclei in the course of clouds formation, contribute to the increase of albedo (Twomey 1974; Roberts et al. 2008). However, due to the multiplicity and complexity of the aerosols interactions, the estimates of climate transformations range due to the aerosols are still charged with a significant error (IPCC 2001, 2007). Among natural constituents of aerosols, those playing a key role are: sea salt, dust from the dry regions (e.g., Sahara), organic substances of marine and terrestrial origin, and sulfur compounds (originating, mainly, from sea water and volcanic eruptions; Andreae 2007). Paleoclimatic reconstructions indicate the volcanic eruptions as the main cause of the climate transformations in the Earth's past (Robock 2000; Gao et al. 2008).

Climate is also determined by the atmosphere composition. To this end the greenhouse gases play the most significant role (IPCC 2001, 2007). About 30 percent of solar radiation reaching the atmosphere is reflected back into the outer space. The remaining 70 percent heats the lower atmosphere and the planetary surface. In order to balance the heat that reaches the Earth, the planetary surface emits infrared radiation (IR) back into the outer space. Part of the IR is absorbed by the greenhouse gases causing the atmosphere temperature increase—the mechanism called a greenhouse effect (Kiehl and Trenberth 1997).

This phenomenon allows to maintain the Earth's temperature at a level that is both high enough and stable to create habitable conditions on the Earth surface. Among the most important greenhouse gases are: water vapour, CO_2, CH_4, N_2O, CO and CFCs. Due to its partial pressure and the absorption spectrum, water vapour plays the key role in the atmosphere, contributing up to 95 percent of the greenhouse effect (Le Treut et al. 2007), while the role of the remaining gases is, at least, one order of magnitude smaller. Nonetheless, paleoclimate studies clearly indicate that global temperature variations are largely related to CO_2 and CH_4 concentrations changes. This conclusion is based on the both gases concentrations in the ice core collected in the Antarctic (Lake Vostok). The deepest layers of the core are dated on 400,000 years BP (Petit et al. 1999), allowing to reconstruct climate of the several glaciations (Monnin et al. 2001). Moreover, as demonstrated by Clark and Mix (2000), significant change in CO_2 concentration was accompanying glacier retreat about 19,000 years ago, strengthening the effect caused by the Sun radiation increase.

2.1.2 Climate Change and Anthropogenic Influences

During the last few decades there has been increasing interest in the range and intensity of anthropogenic activities influencing the environment. Worldwide impact of socio-economic development resulting in the climate alteration has been the main concern of the major scientific projects, e.g., JGOFS (Karl et al. 2001), IGBP (IGAC 2006), CARBOEUROPE, and CARBOOCEAN (Schulze et al. 2009), as well as a reason for the political concern (e.g., IPCC 2007).

The prerequisite for the initiation of climate change is a shift in the Earth's energy balance that may be a consequence of at least one of the following mechanisms (Le Treut et al. 2007):

- Change in the amount of solar radiation reaching the Earth's surface—mechanism independent of anthropogenic activity, as described in Sect. 2.1.1.
- Change in the ratio of incident to reflected solar radiation (albedo)—mechanism depending on both anthropogenic activity and naturally occurring events in the environment.
- Change in the IR radiant flux being re-emitted by the surface—mechanism depending on anthropogenic factors (greenhouse gases concentration in the atmosphere).

The surface covered by ice (cryosphere) has a great impact on the Earth's albedo (Groisman et al. 1994). For example, solar radiation is much more efficiently reflected from a surface of ice, characterized by albedo of 0.8, than from other regions of the Earth, with albedo averaging at around 0.3.

The low average albedo of the Earth's surface is determined primarily by a large surface area and low albedo (<0.1) of the ocean (Le Treut et al. 2007). Recent studies demonstrate that the cryosphere surface has been reduced in size with simultaneous weight loss of the ice, estimated at 7,000 Gt (Gt $= 10^9$ tons) in the period 1960–2003. Clark et al. (1999) demonstrated that snow and ice melting rates are significantly enhanced by positive feedback from the albedo decrease. An additional factor lowering the cryosphere's albedo, particularly in the Northern Hemisphere region, is caused by the increasing amount of soot particles. Among the important sources of these airborne contaminants are the fossil fuel and biomass burning (Hansen and Nazarenko 2004; Jacobson 2004). On the other hand, the Earth's albedo is enhanced by aerosols produced by humans, mainly in combustion processes (Fig. 2.3; Penner et al. 2001, 2003, 2004). A common criterion used to differentiate between aerosols is their chemical composition (Forster et al. 2007):

- Sulphate aerosols—originating mostly from fossil fuels combustion processes (72%), while biomass combustion contributes merely 2% of the total. The other important aerosols are: dimethyl sulfide (DMS) produced by phytoplankton, amounting to 19% of the total; and sulphites formed during volcanic activity, amounting to 7% of the total (Penner et al. 2001).
- Organic aerosols—formed during combustion processes of fossil fuel, biofuels, and burning of forest areas. Ito and Penner (2005) estimated that about 2.2 Tg (Tg $= 10^{12}$ g) of particulate organic matter (POM) is emitted annually from fossil fuels combustion and 7.5 Tg from biofuels combustion. The present-day rates for organic aerosols emission are three orders of magnitude higher than in 1870.

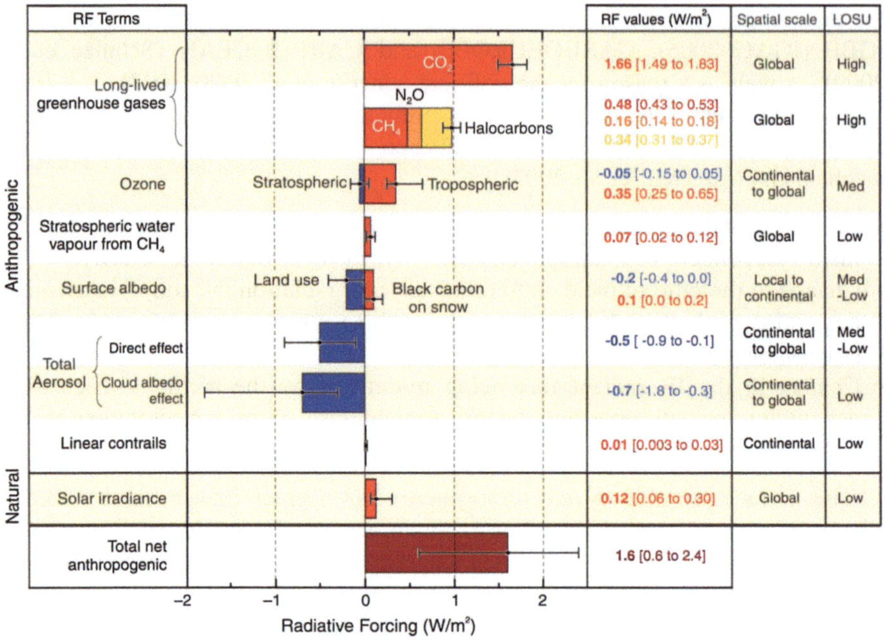

Fig. 2.3 Global radiation forcing for specific constituents in 2005, in relation to the year 1750, together with levels of scientific understanding (LOSU; IPCC 2007)

- Nitrate aerosols—their formation depends on substantial concentrations of ammonia and nitric oxide in the atmosphere. The nitrate aerosols are of lesser significance than sulphate aerosols (Schaap et al. 2004).
- Mineral dust—derived from naturally occurring processes (e.g., Sahara); 30–50% of the dust originates from human activity such as agriculture, cement production and melioration (Tegen and Fung 1995). Estimates regarding mineral dust contribution in the energy balance are contradictory (Forster et al. 2007).
- Carbon soot—unlike the aerosols described above, it is characterized by a positive radiation forcing.

The overall effect of aerosols originating from the anthropogenic activity on the Earth's albedo is additionally strengthened by changes occurring on the land (Fig. 2.3). Deforestation, forests burning and arable areas expansion are main processes contributing to the changes (Forster et al. 2007). Increase in the Earth's albedo caused by development in agriculture is, on the other hand, especially pronounced in mid-latitude regions, where during the winter period meadows and arable areas are coved by snow more expansively than forest areas. Klein Goldewijk (2001) suggest that the actual meadow and arable areas expanded from 7.9–9.2 mln km^2 (6–7% of the total land area) in 1750 to 45.7–51.3 mln km^2 (35–39%) in 1990, with simultaneous decrease in total forest areas by 11 mln km^2.

However, the estimate of the effects of the above transformations and their influence on the Earth's albedo is still charged with a considerable error (Fig. 2.3; IPCC 2007; Forster et al. 2007).

In contrast to aerosols, characterized by the greatest negative radiation forcing, the greenhouse gases are a key element leading to an increase in global temperatures (Fig. 2.3; IPCC 2007). The impact of anthropogenic greenhouse gases began around 8,000 years ago (Ruddiman 2005). Starting from then the natural variability of CO_2 concentration in the atmosphere has no longer correlated with Milankovitch cycles (natural changes in the Earth's surface insolation; Bates 1987). Assumed cause of this change, that occurred 8,000 years ago, is the beginning of agricultural settlements in the Northern Hemisphere. Deforestation had been leading to increased CO_2 emission to the atmosphere, decreasing at the same time the natural absorptive capacity of the biosphere in relation to CO_2 (Ruddiman 2003). In the case of CH_4, a similar divergence dates back to about 5,000 years ago, related to extensive agricultural practices, mainly rice production in South-East Asia (Ruddiman 2003).

According to IPCC (2007) report, the major greenhouse gases entering the atmosphere which result from human activity include: CO_2, CH_4, N_2O and freons (Fig. 2.3). In particular, concentration of freons increased significantly during the 1990s, as a result of widespread use of freons in refrigerators, cosmetics production and pressurized polymers (Velders et al. 2005). Although the freon and related compounds concentrations in the atmosphere are relatively small (approx. 10^6 times lower than CO_2), their high capacity to absorb the infrared radiation contributes significantly to the greenhouse effect (Fig. 2.3; Forster et al. 2007).

The relative increase in concentration of methane (CH_4) in the atmosphere began in 1750 (Fig. 2.4; Forster et al. 2007). This greenhouse gas is characterized by the highest after CO_2 radiation forcing. The most important anthropogenic sources of CH_4 in the atmosphere include agriculture, especially rice cultivation, cattle breeding, landfills and leaks in systems extracting and transporting natural gas (Forster et al. 2007).

Regular monitoring of methane concentrations in the atmosphere indicates significantly reduced growth rate and stabilization of methane during recent years (Dlugokencky et al. 2003). However, the mechanisms of these changes are not fully understood (Forster et al. 2007).

Among all the greenhouse gases, CO_2, that most significantly contributes to an increase in the global temperature, deserves a special attention (IPCC 2007). Radiative forcing of CO_2 in 2005 compared to 1750 exceeds the sum of radiative forcing of all the remaining greenhouse gases (Fig. 2.3). In addition to the natural sources of CO_2 in the atmosphere, that are associated mainly with the organic matter cycle, human introduces even more CO_2, which further disrupts the environmental balance, leading to the climate warming. The main sources of the anthropogenic CO_2 include: the fossil fuels combustion (Forster et al. 2007), cement production (Worrell et al. 2001), deforestation—especially tropical forests (Houghton 2003; Strassmann et al. 2008), and—to a smaller extent—the biomass combustion (Andreae and Merlet 2001).

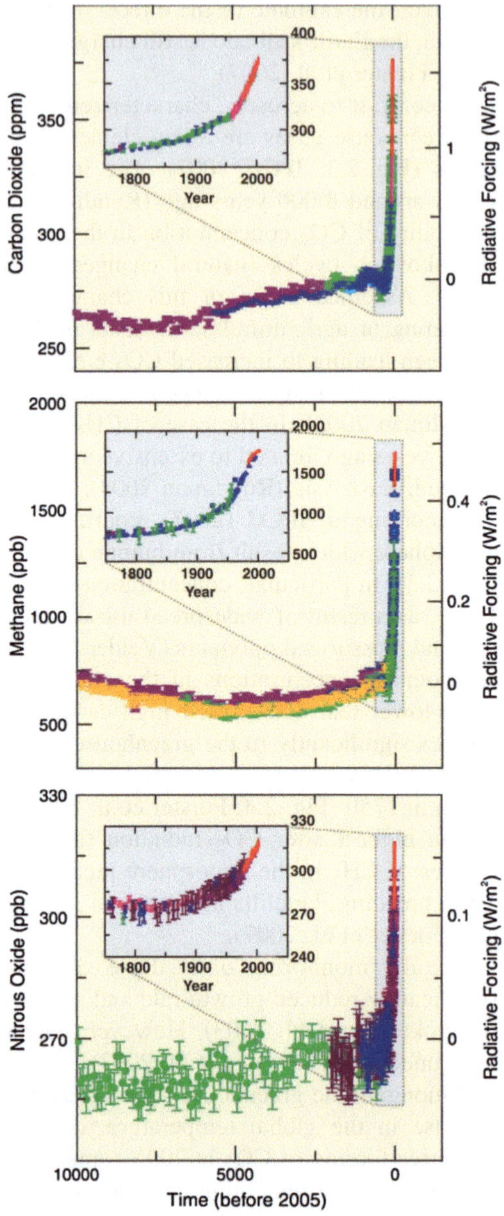

Fig. 2.4 Concentrations of atmospheric CO_2, CH_4 and N_2O during the last 10,000 years, reconstructed from ice cores. Different colors indicate different sources of data. Results of recent measurements are in red. Radiative forcing is calculated in relation to the year 1750 (IPCC 2007)

Intensification of these processes, especially in recent decades, in particular the fossil fuels combustion, has led to a significant increase in the atmospheric CO_2 concentration (Fig. 2.5; Keeling et al. 1995). The monitoring studies initiated in the late 1950s (Fig. 2.5) suggest large seasonal variations of CO_2 in the atmosphere, caused by heterogeneous emission of anthropogenic CO_2 during the year,

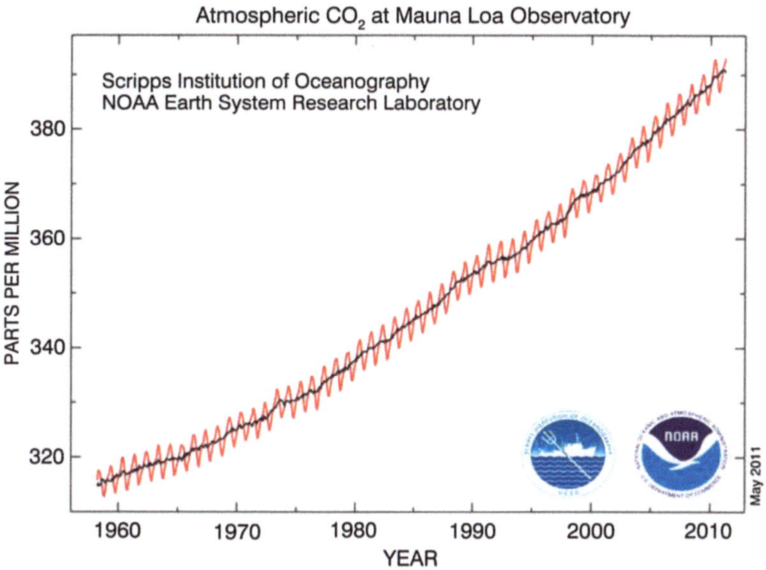

Fig. 2.5 Changes of the atmospheric CO_2 over decades; observatory Mauna Loa, Hawaii (source: http://www.esrl.noaa.gov)

as well as significant contribution of naturally occurring processes contributing to the global carbon cycle. Therefore, research that leads to understanding of the mechanisms controlling CO_2 concentration in the atmosphere should be in the center of today's earth sciences interests. Uncertainty regarding the magnitude of anthropogenic CO_2 emission in the future, due to the socio-political and economic interactions of a modern world, gives a further motivation to take actions (IPCC 2007).

2.1.3 Global Consequences of the Climate Change

It is assumed, with a high certainty, that human activity, primarily the greenhouse gases emission, has led to a global temperature increase (IPCC 2007). It is also believed, although with limited certainty, that these actions prevented a new ice age to begin from 3,000 to 4,000 years ago (Ruddiman 2003; Ruddiman et al. 2005). According to the last IPCC report (IPCC 2007), the measurable increase of the global temperature during the last fifty years is a direct evidence of human interference with the atmosphere composition. Models that are taking into account only natural radiative forcing indicated clearly the inevitability of cooling for this period of time. Studies conducted by Trenberth et al. (2007) show that the average global temperature for the period from 1906 to 2005 increased by $0.74 \pm 0.18°C$, while the temperature increase during the last fifty years has been two times faster than during the entire century and amounted to 0.013°C $year^{-1}$. This value is relatively small, mainly due to the heat exchange through vast ocean surface, which moderates the temperature changes. Above the continents, the temperature has been rising since 1979 at a rate of 0.027°C $year^{-1}$. Furthermore, the temperature increase varies in different parts of the world, and in some regions (Arctic and Antarctic) is subject to large fluctuations during subsequent decades (IPCC 2007; Trenberth et al. 2007).

Rapid, from the viewpoint of geological time scale, temperature increase carries numerous consequences, which not only disturb the environment as a whole and its compartments, but also force changes in mankind development. One of the major effects of the atmospheric temperature rise is melting of mountain glaciers, ice sheets and a decrease of both the sea ice surface and volume (IPCC 2007). This is due to impaired annual balance of ice growth and diminishing. High rates of this phenomenon allow to track the process of an ice regression over time—as short as one generation span (Georges 2004; Klein and Kincaid 2006; Gordon et al. 2008). Drinking water resources shrinkage is one of the immediate results of melting glaciers. Approximately 1/6 of the globe population live within or near river basins supplied by water fed from melting glaciers and periodically occurring snow (Kundzewicz et al. 2007). Diminishing glaciers have also their socio-economical aspect due to employment rate of the local communities, often relying on tourism and winter sports (Breiling and Charamza 1999). At a global scale, melting of glaciers influences the air masses circulation, as demonstrated by changes taking place in the Himalaya Mountains (Lemke et al. 2007), and, to some extent, sea level (IPCC 2007).

The cryosphere melting is a direct cause of rising sea level (Hagen et al. 2003; Alley et al. 2005). It is estimated that in the period from 1961 to 2004 the water from melting glaciers has been increasing the sea level in a rate of 0.5 ± 0.18 mm $year^{-1}$. During recent years this process has become even more pronounced, and the rate of melting, expressed as the raising sea level, increased to 0.77 ± 0.22 mm $year^{-1}$ (Lemke et al. 2007). In the case of ice sheets, diagram of long term trends is less precise, because their volume is subjected to large fluctuations over decades (IPCC 2007). However, recent studies (Lemke et al. 2007) indicate that the melting ice sheet of Greenland contributed from 0.14 to 0.28 mm $year^{-1}$ to a rate of sea level increase, during the period 1993–2003. Estimates for Antarctica are less precise and oscillate, for the same period of time, between 0.55 mm $year^{-1}$ (increase in sea level) and −0.14 mm $year^{-1}$ (decrease).

The total contribution of cryosphere to a global sea level increase for the years 1993–2003 is estimated at 1.2 ± 0.4 mm $year^{-1}$ (Lemke et al. 2007). The risk carried by the melting ice is stressed by the fact that the water volume stored in the ice sheet of Greenland and Antarctica, expressed as an equivalent of sea level rise, amounts to about 64 m (Lythe et al. 2001; Lemke et al. 2007). However, no one expects that the sea level rise will exceed 1–2 m by the end of this century.

Thermal expansion of water also influences the sea level (Miller and Douglas 2004). Bindoff et al. (2007) estimate that in the period 1961–2003 the global average temperature of the oceans increased by 0.1°C. This contributed to the sea level increase at a rate of 0.4 ± 0.1 mm $year^{-1}$. Similarly to the rate of melting glaciers, the influence of water thermal expansion on the sea level increase has intensified over the past several decades. The value for the period from 1993 to 2003 has been estimated to be 1.6 ± 0.5 mm $year^{-1}$. When summed up with the cryosphere melting it gives the sea level increase rate of 2.8 ± 0.7 mm $year^{-1}$ (IPCC 2007).

Results of paleoclimatic reconstructions suggest a natural oscillation of the sea level reaching as much as 120 m (Fairbanks 1989). However, during the last 3,000 years or more, the sea level was stable. When the recent trends continue, a risk of local, and regional, flooding of low-laying areas will materialize in the nearest future (Bindoff et al. 2007; Nicholls et al. 2007).

Another consequence of elevated concentration of CO_2 in the atmosphere, related directly to the oceans, is sea water acidification (Fig. 2.6; Caldeira and Wickett 2003; IPCC 2007). The sea water pH is determined primarily by the so-called carbonate buffering system—the presence of ions and, simultaneously, of dissolved CO_2. The increase of CO_2 in the atmosphere results in the increase of partial pressures difference of the gas between the atmosphere and the ocean. This, in turn, leads to the diffusion of CO_2 into water (mechanism described in some detail in Sect. 2.2.3.2). In this way, the present day carbonate equilibrium is disturbed resulting in pH lowering. This is a particular threat for the marine organisms that built their shells out of $CaCO_3$, whose stability depends directly on the pH of water (Feely et al. 2004; Orr et al. 2005). The gradual drop of the seawater pH has amounted to 0.1 pH unit (Fig. 2.6). Direct measurements indicate that the rate of pH drop increased in the last twenty years to 0.02 pH units per decade (Bindoff et al. 2007).

The climatic changes bring also an increasing number of extreme events (Kundzewicz et al. 2004; Trenberth et al. 2007) recognized as 'low frequency of occurrence events'. These are often characterized by high intensity, and pronounced effects, e.g., erosion, material losses, storms and storm surges (BACC Author Team 2008). Trenberth et al. (2007) indicate at the increase in intensity and frequency of occurrence of tropical cyclones, hurricanes, typhoons, being related directly to an increase of the surface water temperature in tropical regions.

An increase of the average global precipitation was also observed during the 20th century. At the same time, the frequency of intense rainfalls has increased (Trenberth et al. 2007), resulting in numerous flooding events (Ulbrich et al. 2003) and droughts, particularly since the 1970s (New et al. 1999; Kundzewicz 2009). Extremely dry weather conditions are more often encountered in tropical and subtropical, as well as mid-latitude regions, e.g., Western Europe. Droughts occur, commonly, along with the unusually high temperatures (Trenberth et al. 2007).

Global material losses attributed to extreme events increased eight times during the last 40 years (inflation included). Proportion of goods insured against the natural disasters has also increased over the recent years. This situation is related partly to the growing wealth and development of societies, but also largely to the growing frequency of the extreme climate events occurrence (Mills 2005; Beniston 2007; Kundzewicz 2009).

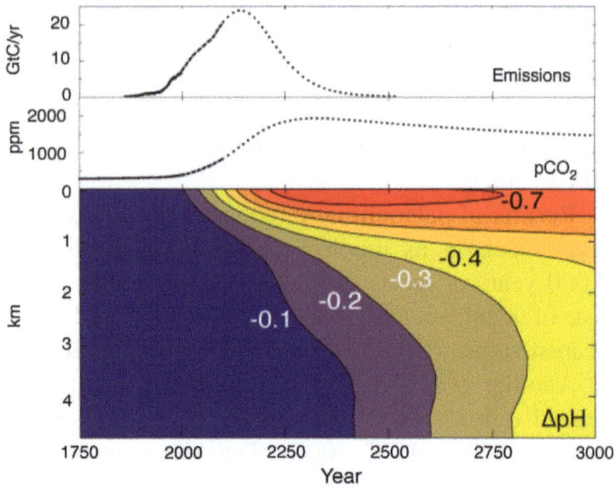

Fig. 2.6 Predicted pH changes in the ocean, based on target values of CO_2 emission and concentration in the atmosphere (Caldeira and Wickett 2003). The *Y* axis represents alternately the extend of CO_2 emission to the atmosphere (*topmost* dependence), concentration of CO_2 in the atmosphere (*middle*) and the pH drop through the depth of the ocean

Climate changes exert pressure on the natural ecosystems. According to Fischlin et al. (2007), approximately 20–30% of flora and fauna species will be endangered before 2100. This is a consequence of rising temperature and the atmospheric CO_2 concentration to levels unpresented during the last 650,000 years for CO_2, and 740,000 years for temperature (Fischlin et al. 2007). Negative changes will also affect marine ecosystems, mostly organisms built of $CaCO_3$, coral reefs being particularly threatened (Riebesell et al. 2007; Wei et al. 2009). The projected increase in thermal stratification for the tropical and mid-latitude ocean zones can lead to reduction of primary production in these regions (Doney 2006).

The transformations that are observed now and projected for the future, and their consequences, provoke the society to act against these perturbations or, at least, to adapt to them. Challenges, that the future generations will be facing include:

- Providing food and drinking water to a constantly growing population, affected by more frequent and intense climate events and shifts of climatic zones (Easterling et al. 2007; Bański 2009);
- Development of systems that warn and protect population against the extreme climate events (Kundzewicz 2009);
- Evoking public awareness about the existing risks and adaptation means (Kundzewicz 2009);
- Protection of the public health against harmful effects of the climate warming (Confalonieri et al. 2007; Błażejczyk 2009);
- Protection of the endangered areas, primarily coastal and polar regions (Nicholls et al. 2007; Anisimov et al. 2007);
- Deepening of the knowledge about mechanisms determining the magnitude of anthropogenic factors influencing the Earth's climate, particularly concerning the global carbon cycle (IGBP 2002; IPCC 2007; Gutry-Korycka 2009; Schulze et al. 2009).

Assuming more conscious human actions and accepting the responsibility for climate changes, preventing transformations becomes crucial, especially in the field of greenhouse gases reduction in the atmosphere (IPCC 2007). This is possible to achieve by optimizing energy generating technologies and developing alternative energy sources (IPCC 2007; Kintisch 2009).

All the above actions involve considerable costs, that very often exceed a single country budget, in particular budgets of developing countries. Thus there is a need for balanced responsibility, shared between the developed and developing countries (IPCC 2007; Kintisch 2009).

2.1.4 Climate Change Consequences in the Baltic Sea Region

The Baltic Sea climate is characterized by a large seasonal variability, typical of its geographical location, topography of the catchment area, and the sea–land interactions (HELCOM 2007). Characterized by western winds dominance, the south-western part of the basin is under a strong influence of the North Atlantic and the Gulf Stream. This effect weakens in the north-eastern direction, where the continental influences and polar air masses dominate. Winter season climate features, for the Baltic, and most of the Europe are determined mainly by NAO (BACC Author Team 2008).

Many of the conditions determining changes within the Baltic Sea region are identical to those that control other parts of the world (BACC Author Team 2008). Against a background of globally increasing temperature, estimated for the period 1861–2000 to raise by 0.05°C $decade^{-1}$, the warming of the Baltic Sea region seems to be actually greater, reaching up to 0.08°C $decade^{-1}$ (HELCOM 2007).

Sea level is quite dynamic in the Baltic. An important factor contributing to the Baltic basin bathymetry is related to isostatic movements of the Earth's crust. It is estimated that, as a result of these movements, the bottom of the Gulf of Bothnia uplifts at a rate of 10 mm per year, whilst the southern part of the Baltic Sea lowers by 0.5–2 mm per year (Wyrzykowski 1985; Harff et al. 2001; Johansson et al. 2004). Changes occurring in the North Sea and the North Atlantic also impact sea level in the Baltic (Pempkowiak et al. 2009). Models suggest that, by the year 2100, the Baltic Sea level will rise in the range from 46 (Danish Straits) to 100 cm (Polish coastal zone), with simultaneous drop from several to several dozen centimetres in the Gulf of Bothnia (Fenger et al. 2001; Miętus et al. 2004; Schmidt-Thomé et al. 2006; Pempkowiak et al. 2009). Such a high increase of the water level poses a threat, especially to low-laying areas, with Polish coast being particularly exposed. Pempkowiak et al. (2009) estimate that as much as 1,800 km^2 of the area along the sea shore could be covered with water (Fig. 2.7). According to Schmidt-Thomé et al. (2006), significant losses could be also observed in the City of Gdansk and within the Vistula Delta.

The effect of rising water levels in the Baltic is enhanced by storms and storm surges (Pempkowiak et al. 2009) and by predictably more frequent occurrence of

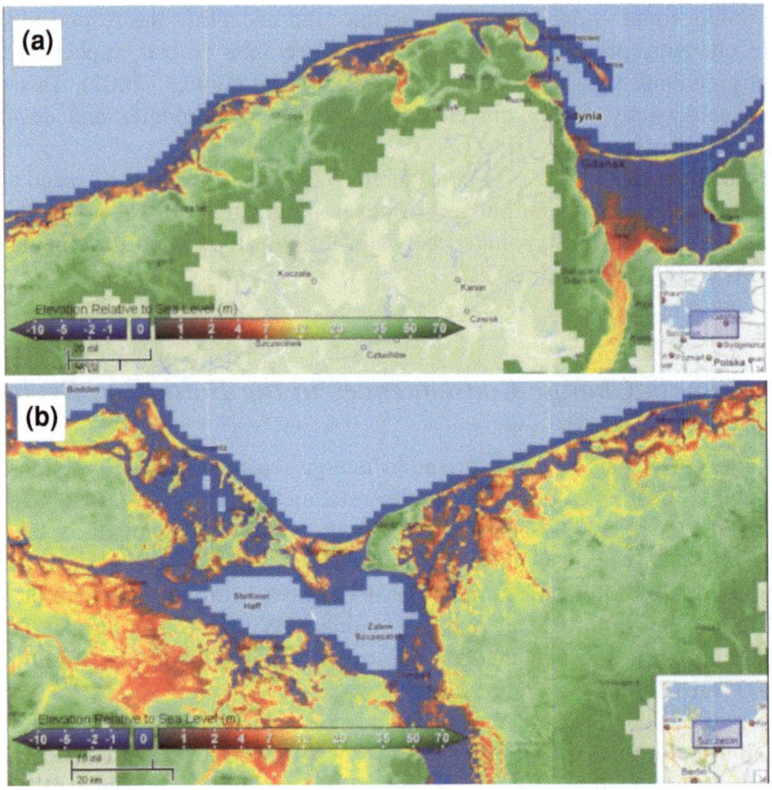

Fig. 2.7 Map of the Southern Baltic coast, threatened by inundation as a result of rising sea level: **a** central and eastern coast; **b** Pomeranian Bay and the Szczecin Lagoon coasts. Colours correspond to the elevation of coastal areas (source: www.globalwarmingart.com)

extremely intensive rainfalls in this part of Europe (HELCOM 2007). These factors pose a considerable threat to the regions situated in vicinity of estuaries; an example may be the unexpected flood that struck Gdańsk, a port city on the southern Baltic coast, in 2001 (Majewski 2003). Storm occurrence frequency is expected to increase during the coming decades, while trends in wave height are not demonstrated (Miętus 1999; Meier et al. 2004; Schmidt-Thomé et al. 2006). Rising sea level, as well as storm events and storm surges escalation, can lead to an enhanced erosion of sea shores. Redeposition coefficient—shore line regeneration to its erosion ratio—is an often used measure to describe erosion intensity. The data compiled by BACC Author Team (2008) suggest that the value of the coefficient for the Polish coastline has been decreasing: for the period 1875–1979 the redeposition coefficient amounted to 69%, between 1960 and 1983 it was reduced to 20%, to reach just 14% during 1971–1983. The combination of the above-mentioned factors creates a risk for the local infrastructure: residential areas, agricultural holdings, tourist resorts, and ports (Liszewska 2004; Pempkowiak et al. 2009).

Climate transformations carry important consequences for the Baltic Sea ecosystem (HELCOM 2007; BACC Author Team 2008). The atmosphere temperature increase leads to reduction of convective mixing, followed by nutrients cycles alterations and anoxic conditions in the water layers below the halocline. Anoxia and hypoxia are of particular importance in the reproductive cycle of cod. Changes in the population of this top Baltic predator result in transformations along the entire food chain (Mackenzie et al. 2007). Temperature increase transforms also the ecosystem structure, by favouring the warm water species over the cold water ones (e.g., cyanobacteria vs. diatoms; Wasmund and Uhlig 2003). Impact of the projected decrease in salinity and water acidification (Stigebrandt and Gustafsson 2003), caused by elevated CO_2 concentration (Omstedt et al. 2009) will additionally accelerate the Baltic Sea ecosystem transformations (HELCOM 2007).

2.2 Global Carbon Cycle

2.2.1 Carbon Reservoirs

Carbon belongs to the most widespread chemical elements on Earth, yet constitutes only 0.0012% ± 0.00036% of its total mass (Lécuyer et al. 2000). There are three carbon isotopes occurring in nature: two stable isotopes, ^{12}C and ^{13}C, and radioactive ^{14}C (half-life about 5,730 years; Sulzman 2007). Despite its low contribution to the earth mass, carbon is fundamental for life on Earth, being the essential element present in all known life forms. Another story, but of great significance, is the previously described greenhouse effect determined, to a great proportion, by carbon compounds.

Data concerning reservoirs and carbon fluxes among the environment compartments have been cataloged and published in numerous recent studies (e.g., Doney et al. 2003; Sarmiento and Gruber 2006; Houghton 2007; Emerson and Hedges 2008; Schulze et al. 2009). The majority of available reports are based on the same source data, and present comparable estimates, with differences seldom exceeding few percent. For the purpose of the present summary, the compilations of data by Emerson and Hedges (2008) and by IPCC (2007) have been used (see Fig. 2.8).

Carbon is present in all the three domains of the natural environment: land, water and atmosphere. The smallest amount of carbon, mainly in the form of CO_2, is present in the atmosphere. It is estimated that the amount of atmospheric carbon dioxide, during the pre-industrial period, was approximately 612 Pg ($Pg = 10^{15}$ g), compared to the current 784 Pg. A comparable amount, of roughly 600 Pg, is typically assigned to terrestrial organisms, with dominating proportion (80%) of carbon occurring in the plants biomass. Additional 1,500 Pg of carbon is gathered in soil. Another reservoir of carbon are the fossil fuels in the Earth's crust, which are currently the most widely used sources of energy. The greatest carbon amount occurs in rocks: coal, lignite, slate and oil shale. It is estimated that these minerals contain up to a total of 3,440 Pg of carbon. Much less carbon is present in oil deposits (about 90 Pg) and natural gas (about 44 Pg). However, the dominant reservoir of carbon is the ocean. The amount of carbon accumulated there is equal to 38,000 Pg, which represents 80% of the total carbon mass in the Earth crust. Furthermore, 1,000 Pg of carbon is deposited in marine sediments. Contribution of marine organisms is less significant (Fig. 2.8; Emerson and Hedges 2008).

2.2.2 Scheme of the Global Carbon Cycle in the Environment

The carbon cycle is a part of the global matter cycle. The amount of carbon comprised in individual reservoirs (Sect. 2.2.1) illustrates their potential impact on the cycle. The principal matter carrier in the carbon cycle is CO_2. In the absence of human influence, carbon cycle is characterized by the steady state (Fig. 2.8). Carbon exchange between individual reservoirs involves four fundamental processes: photosynthesis, respiration, gas exchange through the water/atmosphere interface and weathering.

Photosynthesis and respiration are the primary processes facilitating carbon exchange between the land and the atmosphere. During the photosynthesis, organisms capable of carbon assimilation, mostly plants and cynobacteria, absorb CO_2, and, with participation of H_2O and solar energy, they synthesize organic compounds forming the organisms' biomass. Animals and microorganisms, as the successive levels of a food chain, utilize the biomass, enabling further carbon cycling. Most of the living organisms oxidize organic matter in order to generate energy necessary for them to function. Besides energy, H_2O and CO_2 are the final products of the oxidation. The resulting CO_2 is most often released to the atmosphere. Carbon dioxide exchange between land and the atmosphere is estimated at 120 Pg C year^{-1} (assuming complete respiration of organic matter on land, without including erosion and, resulting from erosion, dissolved carbon species runoff to the ocean; Fig. 2.8; Emerson and Hedges 2008). In a global scale, erosion (approximately 0.4 Pg year^{-1}) does not play a significant role (Schulze et al. 2009). Somewhat smaller amounts of carbon are subject to the exchange through the water/atmosphere interface, estimated at 90 Pg C year^{-1} (Fig. 2.8; Emerson and Hedges 2008). Processes of absorption/emission of CO_2 by seas and oceans, and carbon fluxes in the marine environment are described, in some details, in Sects. 2.2.3 and 2.2.4.

Weathering involves physical and chemical processes. The latter, named chemical weathering, is caused by the atmospheric CO_2, that, combined with H_2O, reacts with carbonate and silicate rocks. The result of weathering are ions, which, together with river water, are discharged to the ocean. It is estimated that two-thirds of this load is represented by ions derived from carbonates, mostly $CaCO_3$, and the remaining one-third from silicates, weathering. The contribution of

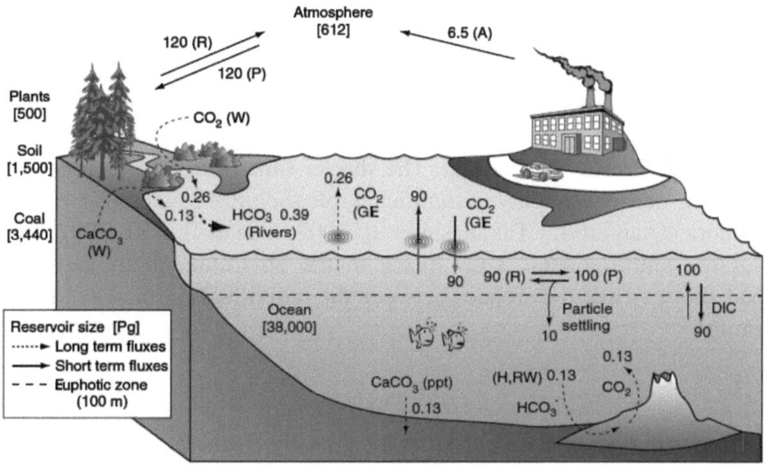

Fig. 2.8 Scheme of the global carbon cycle (Emerson and Hedges 2008). Data expressed in Pg [10^{15} g]. Carbon resources for individual components are marked by *square brackets*. *Dashed arrows* indicate long-term fluxes, *solid arrows*—short term fluxes. Symbols: *W* weathering, *GE* gas exchange through the water/atmosphere interface, *P* photosynthesis, *R* respiration (oxidation), *H* hydrothermal processes, *RW* rock weathering

weathering processes to the global carbon fluxes is negligible, since the total carbon flux related to weathering amounts to, roughly, 0.39 Pg C year^{-1} (Emerson and Hedges 2008).

The steady state global carbon cycle had been disturbed by human activity (Sect. 2.1.2). According to the 4th IPCC report (IPCC 2007), the anthropogenic CO_2 emission to the atmosphere has been increasing yearly. In 2004 it was estimated at 10.4 Pg C year^{-1}, being remarkably higher (by 80%) than the emission recorded in 1970. This represents more than 8% of the natural CO_2 exchange between land and sea, and over 11% of exchange between the ocean and the atmosphere (Emerson and Hedges 2008). Estimated that, in the period from 1800 to 1994, nearly 244 Pg C reached the atmosphere as a result of anthropogenic activity. This is close to twice the amount of known carbon resources stored as crude oil and natural gas, however less than 10% of carbon reserves deposited as coal (Sect. 2.2.1).

Still, the existing fossil fuel deposits, and possible scenarios of the socio-economic development, suggest that by 2030 the emission of anthropogenic CO_2 to the atmosphere can amount to between 140% and 210% of the current levels. Thus, the antrophogenic contribution is likely to be comparable to the natural processes determining CO_2 fluxes in the environment (IPCC 2007).

Quantitative assessment of the anthropogenic CO_2 releases refers to the gas emission during the 1990s (IPCC 2001, 2007). The data demonstrate that as a result of combined fossil fuels combustion and cement production, 6.3 ± 0.4 Pg C year^{-1} reached the atmosphere, out of which close to 3.2 ± 0.1 Pg C year^{-1} (50%) had been retained in the atmosphere. The remaining 3.1 Pg C year^{-1} had been absorbed by the land plants (1.4 ± 0.7 Pg C year^{-1}) and oceans (1.7 ± 0.5 Pg C year^{-1}; Sabine et al. 2004; Emerson and Hedges 2008). The smaller contribution of the continents (despite greater natural CO_2 exchange with the atmosphere compared to the ocean; Fig. 2.8.) is explained by the shrinkage of intensive photosynthesis areas, mainly tropical forests (Sect. 2.1.2). The shorter time of carbon residence on land than that in the ocean, also plays a significant role (Schulze et al. 2009). Estimates by other authors (Quay 2002; Takahashi et al. 2002; Canadell et al. 2007; Houghton 2007) of the anthropogenic CO_2 fluxes in the environment are close to these described above, while differences between the individual estimates fall within the error limits.

2.2.3 Carbon Cycle in the Marine Environment

Significant accumulation of carbon loads in the ocean (Sect. 2.2.1) and the ocean significant contribution to the anthropogenic CO_2 absorption (Takahashi et al. 2002) intensified studies that led to advances in the marine environment carbon cycle, especially during the last decade (Sarmiento et al. 1998; Stein and Macdonald 2004; Sarmiento and Gruber 2006; Emerson and Hedges 2008). In order to access the role of the seas and oceans in the global carbon cycle, complex biogeochemical models are designed that are supposed to reflect mathematically the processes occurring in marine systems (Maier-Reimer and Hasselmann 1987; Siegenthaler and Joos 1992; Doney et al. 2003; Sasai and Ikeda 2003; Watson and Orr 2003; IPCC 2007). The models are largely based on the experimental studies conducted during cruises of research vessels, and data collected using semi-automated instruments installed on different types of buoys, platforms and commercial vessels (Bates et al. 2000; Sayles and Eck 2009).

2.2.3.1 Carbon Concentration and Forms in Seas and Oceans

Carbon in the marine environment occurs in a variety of forms, from ions characterized by a small molecular weight to large particles suspended in water column. Filtration is the method commonly used to separate the suspended fraction from the dissolved carbon. In the oceanographical studies, the most commonly used are glass fiber filters and cellulose membranes with the pore size ranging from 0.2 to 1 μm, while the commonly adopted boundary between dissolved and particulate species is 0.45 μm. This allows for separation of zooplankton, phytoplankton, majority of bacteria and detritus as particulate matter. The filtrate comprises: viruses, colloids, and dissolved substances (Fig. 2.9; Hedges 2002; Emerson and Hedges 2008).

The criterion used to differentiate carbon species in seawater is the division into organic and inorganic carbon fractions. These in turn are divided according to their

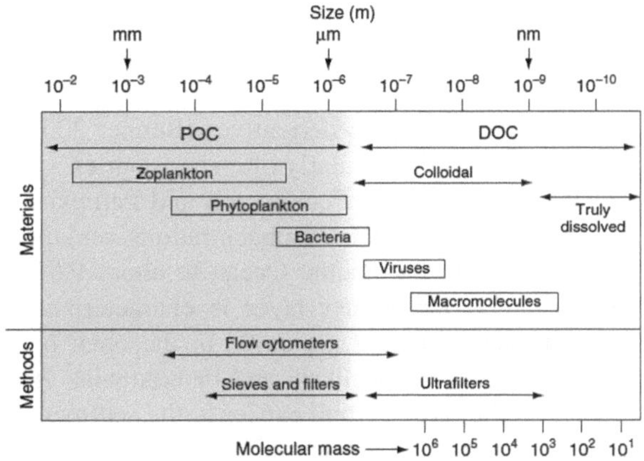

Fig. 2.9 Size of carbon containing components of seawater and methods used to separate and differentiate between the components (Emerson and Hedges 2008)

properties, origin and function in the environment. There are four basic forms of carbon in seawater:

- Dissolved inorganic carbon (DIC).
- Particulate inorganic carbon (PIC).
- Dissolved organic carbon (DOC).
- Particulate organic carbon (POC).

In the marine environment, DIC, as the most abundant carbon species, plays the dominant role in the carbon cycle. Moreover, estimated at 98% of the total carbon that occurs in water, DIC represents the largest resource of carbon globally (Emerson and Hedges 2008). In surface waters, DIC concentrations are subjected to seasonal and spatial oscillations as a result of assimilation and respiration by living organism and CO_2 transport through the water/atmosphere interface. Average DIC concentrations in the surface water layer range from 25–27 mg dm^{-3} for ocean water, to 16–18 mg dm^{-3} for some low salinity estuaries (Thomas and Schneider 1999; Key et al. 2004). The DIC forms include: dissolved CO_2, present in equilibrium with unstable H_2CO_3 that, in turn, is in equilibrium with bicarbonate and carbonate ions: HCO_3^- and CO_3^{2}. These components form the so-called carbonate buffer (weak acid and its salts with strong bases). Carbonate buffer balance is described with dissociation and decay constants of H_2CO_3 (Sarmiento and Gruber 2006). Quantitative ratio of the buffer components concentrations determines pH of sea water, establishing its value at 8.2 at 35 PSU, temperature 20°C and CO_2 partial pressure equal to 360 ppm (Emerson and Hedges 2008).

PIC consist primarily of shells and skeletons of organisms built of $CaCO_3$. After decay of organic matter derived from the dead organisms, settling shells constitue a primary inorganic carbon carrier to the sediments (Godoi et al. 2009). However, the PIC contribution to the marine carbon pool is negligible, and its fraction is rarely distinguished for the purpose of geochemical studies (Milliman 1993; Emerson and Hedges 2008).

DOC is the measure of dissolved organic matter, and also the biggest reservoir of organic carbon on Earth (Hedges 2002). DOC consists of organic carbon in organic substances, e.g., amino acids, carbohydrates, nucleic fatty acid derivatives, humic acids, and lignin derivatives (Benner 2002). The last two groups of compounds are found particularly often in estuaries and land locked seas (Pempkowiak and Pocklington 1983; Pocklington and Pempkowiak 1984; Hedges et al. 1997). In the deep ocean, DOC concentrations remain at stable levels ranging from 0.4 mg dm^{-3} in the Pacific Ocean to about 0.6 mg dm^{-3} in the North Atlantic. The ocean surface water layer is characterized by larger variability of DOC concentrations: 0.6–1.2 mg dm^{-3} in the open ocean and 1–6 mg dm^{-3} in shelf seas (Hansell 2002; Kuliński and Pempkowiak 2008).

POC is the basic carbon carrier to the sediments (de Haas et al. 2002; Emerson and Hedges 2008). Concentrations of POC are determined mainly by the occurrence of phytoplankton and detritus, representing the largest input to the suspended organic matter. Zooplankton and bacteria contribution is much less significant (Andersson and Rudehäll 1993; Stein and Macdonald 2004). POC concentrations in sea water are some 10–20 times lower than DOC concentrations (Hansell 2002; Gardner et al. 2006), and are subject to a large spatial and seasonal variability. Concentrations, in surface waters, range from 0.03 mg dm^{-3} for oligotrophic conditions during winter season, to 0.2–0.8 mg dm^{-3} during summer season in shelf seas, characterized by high primary productivity (Gardner et al. 2006).

2.2.3.2 Factors Determining Carbon Cycle in the Marine Environment

The carbon cycle in the marine environment is determined by a number of physical, chemical, and biological factors that are cross-related. To a large extent, the cycle is controlled by phytoplankton activity (Dzierzbicka-Głowacka et al. 2010; Kuliński et al. 2011). Absorption and assimilation of dissolved CO_2 by phytoplankton leads to disequilibria in the gas partial pressures between seawater (lower) and the atmosphere (higher), causing CO_2 flux into water. On the other hand, in the course of respiration, organic substances of dead biota are oxidized to CO_2. Following the organisms death, the organic matter becomes partially mineralized in the water column and sediments, in effect balancing the loss of CO_2 due to photosynthesis. The so-called biological pump refers to a series of processes including the atmospheric CO_2 absorption by water, CO_2 assimilation leading to organic substances that form biomass, POC mineralization in deeper water layers, and deposition of PIC and non-mineralized POC to the sediments (Fig. 2.10; Longhurst and Harrison 1989; Chisholm 2000; Honjo et al. 2008). In some regions, biological pump is strengthened by upwellings, thermohaline circulation and deep water formation. As a result of these processes, dissolved CO_2 is transported to deeper layers of the water column. Because solubility of CO_2 increases with decreasing temperature, while CO_2 partial pressure in water decreases due to high biological productivity, the North Atlantic becomes the World Ocean region that absorbs the atmospheric CO_2 most efficiently (Takahashi

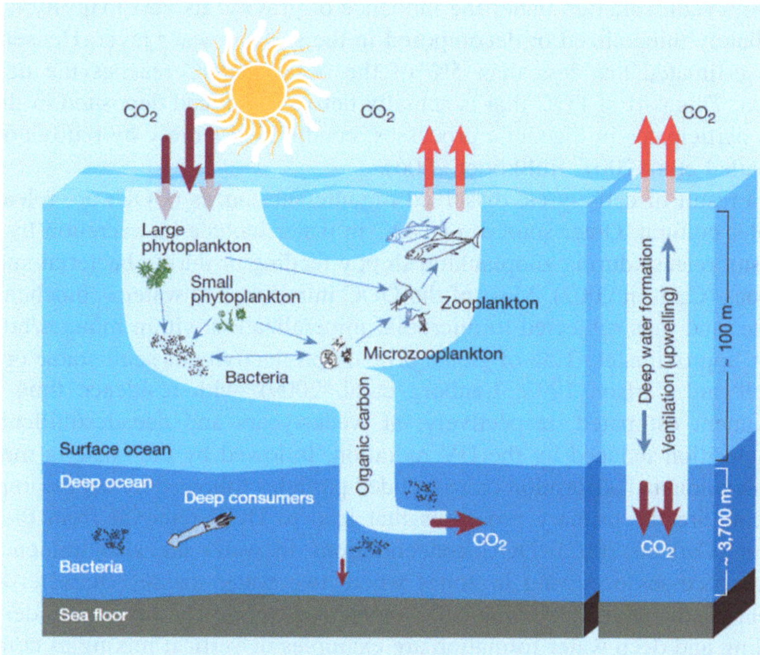

Fig. 2.10 Scheme of the biological pump processes (Chisholm 2000)

et al. 2002; Schuster et al. 2009). In a situation when CO_2 partial pressure is higher in seawater than in the atmosphere, a reverse process takes place, in which the gas flows from water to the atmosphere. Such a situation is observed when respiration (CO_2 production) dominate over the photosynthesis (CO_2 assimilation). Upwelling zones are an example where deep water is brought up to the surface with simultaneous CO_2 degassing to the atmosphere (Torres and Ampuero 2009).

An effectiveness of the biological pump depends largely on physical factors. Since solubility of the majority of gases decreases with increasing water temperature, the CO_2 exchange between water and the atmosphere is determined by the water temperature (Takahashi et al. 2002). It has been demonstrated that the velocity of CO_2 transport through the seawater/atmosphere interface is proportional to the cube of the wind speed (Wanninkhof and McGillis 1999).

The primary productivity—process influencing, at the same time, CO_2 and DIC concentrations in a sea water—also determines organic carbon concentrations. Suspended organic matter concentrations (POC) depend directly on primary productivity (Hedges 2002). On the other hand, the intensity of primary productivity is conditioned by microelements and macronutrients dissolved in water, e.g., nutrients and iron, that are essential for organisms to function. Other factors limiting the phytoplankton growth are solar radiation and water temperature (Moore et al. 2002; Colijn and Cadée 2003; Hiscock and Millero 2002; Noiri et al. 2005). After organisms death, lyses of cells occurs and the resulting suspended organic matter undergoes sedimentation under the influence of gravity. Its vast majority is being immediately mineralized or decomposed in the surface water layer. Hensen et al. (2006) estimated that less than 5% of the oceanic POC reaches the depth of 1,000 m. This part of POC that is actually delivered to, and deposited in the sediments participates in digenetic processes, conditioned largely by redox potential (Frudenthal et al. 2001, Rullkötter 2006).

As a result of cells lyses, dissolved organic substances (DOC) are released to the water column. Other sources of DOC in water include its excretion by living organism, release during zooplankton sloppy feeding or during bacterial and viral infections (Carlson 2002). Most of the DOC introduced to water is biochemically unstable, and thus subjected to microbial mineralization within minutes/hours. A portion, approximately 17% of global primary productivity, is semi-stable or stable (Hansell and Carlson 1998; Lønborg et al. 2009), with residence time in the environment estimated, respectively, at weeks/years and decades/millenniums. Decomposition induced by the UV radiation, followed by the bacteria mediated mineralization and adsorption on suspended particles followed by deposition to the sediments are the primary processes that lead to DOC removal from the environment (Carlson 2002). DOC concentrations in water are also influenced by mixing processes occurring in zones where two water masses characterized by different chemical and/or biological compositions meet. The previously described upwelling and deep water formation are examples of vertical mixing as compared with horizontal mixing that includes events like wind currents, inflow of river water or water exchange between the ocean and shelf seas (Rachold et al. 2004; Thomas et al. 2005; Huertas et al. 2009).

2.2.4 Role of Shelf Seas in the Atmospheric Carbon Dioxide Absorption

Shelf seas are coastal waters connecting the land, the atmosphere and the open ocean; their extent is usually limited to water shallower than 200 m (Thomas et al. 2009). Although the shelf seas represent about 7% of the total surface of seas and oceans, and only 0.5% of their total water volume (Chen and Borges 2009), they play an important role in the global carbon cycle. It is assumed that 20% of total organic matter originating in the marine environment is produced within shelf seas, while 80% of the total organic matter load deposited to the World Ocean sediments is deposited in the shelf seas (Borges 2005).

Significant primary production in the shelf seas is induced by high concentrations of nutrients that reach the sea with rivers draining urbanized or agricultural areas (Gattuso et al. 1998). That is why the shallower seas are believed to be regions where the biological pump processes absorbing the atmospheric CO_2 are more efficient than in the open ocean (Sect. 2.2.3.2; Chen and Borges 2009). Nonetheless, shelf seas are often neglected in the global biogeochemical models that estimate the role of marine environment in the anthropogenic CO_2 cycling

(Sabine et al. 2004). This is due to a large variability of biological, chemical and hydrological properties differentiating not only individual basins, but single shelf sea ecosystems as well. One of the examples is a difference in CO_2 partial pressure of water and air. Its seasonal variation oscillates from about 40 ppm in oceanic regions to 400 ppm in shelf seas of comparable latitude (Thomas and Schneider 1999; Takahashi et al. 2002; Thomas et al. 2004). Assessment of the shelf seas role of in global CO_2 cycling is a serious problem due to a limited spatial resolution of mathematical models (Bozec et al. 2006).

Tsunogai et al. (1999) presented the first available estimate for the role of shelf seas in the global cycle of anthropogenic CO_2. The authors assumed the same, as for the South China Sea, absorption capacity value for all shelf seas. Results proved that shelf seas absorb a total of 1.0 Pg C year^{-1}. Similar estimation technique was used by Thomas et al. (2004), who treated the North Sea as the global standard of CO_2 transport through the seawater/atmosphere interface. The obtained net CO_2 absorption was equal to 0.4 Pg C year^{-1}. The above values emphasize the importance of shelf seas, regardless of their relatively small areas, as regions absorbing carbon dioxide to the extent comparable to the ocean: 1.6 Pg C year^{-1} (Takahashi et al. 2002), 1.7 Pg C year^{-1} (Sabine et al. 2004), 1.9 Pg C year $^{-1}$ (Sarmiento and Gruber 2006).

However, in recent years there have been expressed numerous skeptical opinions regarding the role that shelf seas actually play in the anthropogenic CO_2 absorption (Andersson and Mackenzie 2004; Cai and Dai 2004; Borges 2005; Borges et al. 2005; Borges et al. 2006; Cai et al. 2006; Chen and Borges 2009). The results demonstrate that, in fact, the open water of shelf seas effectively absorbs CO_2 in the amount of 0.2–0.4 Pg C year^{-1} (Borges et al. 2006; Chen and Borges 2009). However, these estimates have not included specific coastal ecosystems: estuaries, upwelling zones, mangrove forests (Chen and Borges 2009). In the meantime, scientific reports from 40 years ago demonstrated that these ecosystems emit CO_2 to the atmosphere (Park et al. 1969; Kelley and Hood 1971). Chen and Borges (2009) estimated this emission at 0.5 Pg C year^{-1}. This load equalizes the CO_2 absorption through the open water of shelf seas. Despite the fact that this result is charged with significant error, it gives a quite new meaning to shelf sea contribution levels in reaching the global balance of the anthropogenic CO_2. The complexity of mechanisms determining functioning of the shelf sea ecosystems results in significant errors of the estimates regarding CO_2 absorption/emission. Insufficient knowledge regarding the matter cycle in shelf seas has been manifested in subsequent data reports, that significantly vary between each other, even if they concern the same areas (Chen et al. 2003; Borges et al. 2005; Chen and Borges 2009).

In the context of the examples cited above, it should be assumed that the Baltic Sea, characterized by intensive phytoplankton activity occurring from spring to fall, additionally supplied by a significant load of nutrients (HELCOM 2009), is a basin that effectively absorbs CO_2. On the other hand, extensive mineralization of organic matter generated during primary production as well as transported by river run off, will contribute significantly to CO_2 production. There is a wide range of literature data available, focusing on the CO_2 exchange between the Baltic and the atmosphere (Thomas and Schneider 1999; Algesten et al. 2004; Kuss et al. 2006; Omstedt et al. 2009). However, all these estimates, reporting local scale studies, still do not provide a clear answer to the question whether the Baltic Sea (as a water body) acts as a CO_2 emitter or absorber, and what is the magnitude of these processes?

2.3 Carbon Cycling in the Baltic Sea

2.3.1 Baltic Sea Characteristics

The Baltic Sea is an inland shelf sea located in Northern Europe, stretching from 53 to 66°N in latitude and from 20 to 26°E in longitude. The basin occupies an area of $3.85 \cdot 10^5$ km^2 (without the Kattegat) and contains a total water volume equal to $22 \cdot 10^3$ km^3 (Fig. 2.11; Gudelis and Jemielianow 1982; HELCOM 2002; BACC Author Team 2008; Lass and Matthäus 2008).

Narrow and shallow Danish Straits (the Great Belt, Little Belt, Oresund—extending through the Skagerrak, and the Kattegat) are the only connection of the Baltic Sea with the North Sea. In a consequence, only infrequent incidents of horizontal water exchange of considerable volume (100–250 km^3) are reported. The water inflows from the North Sea are the cause of density stratification, as well as salinity gradient occurring from the southwest to the northeast. Surface water salinity ranges from 7–8 PSU in the Arkona Deep (south-western part of the Baltic) to 2–3 PSU in the northern part of the Baltic (the Bothnian Bay). The Baltic Sea is considered to be one of the largest brackish water basins in the world. Regardless of its shallowness, inflows of saline and dense seawater from the North Sea is one of the main causes of permanent stratification in regions of the Baltic with water depth exceeding 60–80 m. The halocline, occurring at a depth of 60–80 m, separates the brackish (6–8 PSU) surface water from deeper water layers (11–15 PSU). Limited vertical mixing leads to oxygen deficits and occurrence of hydrogen sulphide. In summer this is enhanced by the thermocline present at 25–30 m and infrequent inflows of the North Sea water, rich in oxygen (Voipio 1981; HELCOM 2002; Stigebrandt 2001; Omstedt et al. 2004; Meier 2007; BACC Author Team 2008; Lass and Matthäus 2008; Matthäus et al. 2008). Significant river runoff estimated at 428 km^3 year^{-1} is the largest source of fresh water in the Baltic, contributing to its brackish character. The volume of precipitation is nearly two times smaller (237 km^3 year^{-1}) and compensated mainly by evaporation estimated at 184 km^3 year^{-1}. The largest river runoff is delivered to the Gulf of Bothnia, the Gulf of Finland, and the Gulf of Riga. Combined the discharges are estimated at 70% of the total volume of fresh water delivered to the sea. The Baltic drainage area is over four times greater than the sea area itself, and covers the territories of fourteen countries. The catchment area is rather diverse geographically—from the mountains and forests of the Scandinavian Peninsula, characterized by a numerous small basins—to

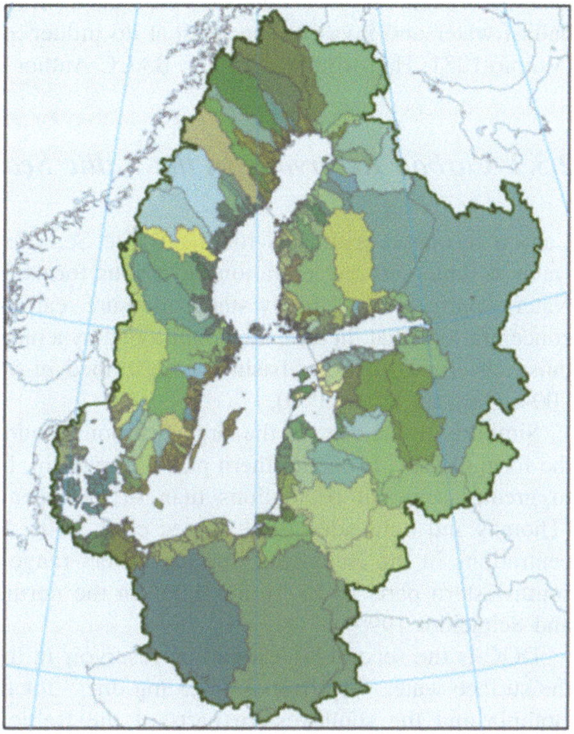

Fig. 2.11 The Baltic Sea drainage area with the highlighted rivers catchments (HELCOM 2007)

lowland mainland areas, transformed agriculturally. The lowland regions, covering the eastern and southern part of the Baltic drainage area, are characterized by large river basins, such as the Neva River, the Vistula River, or the Odra River (Fig. 2.11; Bergström et al. 2001; Omstedt and Axell 2003; Omstedt et al. 2004; HELCOM 2007; Meier 2007; Lass and Matthäus 2008).

The Baltic Sea is characterized by a low salinity that, coupled with the osmotic stress, leads to the ecosystem biodiversity loss expressed in decreasing number of species and the size of individual organisms. On the other hand, Baltic as a transitional area between sub-arctic conditions prevailing in the Gulf of Bothnia and boreal conditions in the southern part of the basin, is a site habitable for species characteristic for both climatic zones. Another factor that significantly shapes the Baltic Sea ecosystem is the human impact. Besides the contaminants introduced to the basin, eutrophication driven by extensive nutrient loads, originating especially from urbanized and agricultural areas, is of particular importance. It is mainly the phosphorus and nitrogen delivered with river runoff that promote the increased ecosystem productivity. In a consequence, intensive mineralization of organic matter occurs, which contributes to a spread out of anaerobic conditions, especially in the deeper water layers of the Baltic. The process of eutrophication greatly promotes the phytoplankton blooms, including toxic cyanobacteria, which additionally becomes a public health concern, especially in the near-shore areas. A separate issue resulting from human activity is a problem of ballast water and invasive species that do influence biodiversity in various ways (Voipio 1981; Hagström et al. 2001; BACC Author Team 2008; HELCOM 2009).

2.3.2 Carbon Reservoirs in the Baltic Sea

Carbon resources estimates for the Baltic Sea are obtained by multiplying the water volume and concentration of various forms of carbon. In a situation when water volume is a relatively stable measure, carbon resources are dependent on concentrations that, in turn, are conditioned by a number of processes analogous to those observed for other basins, as described in Sect. 2.2.3.2 (e.g., Vichi et al. 2004; Omstedt et al. 2009).

Similarly to the oceans, the largest carbon resources in the Baltic are present in the form of DIC. In the northern part of the basin, DIC concentrations are subject to greater seasonal oscillations than in the open ocean or typical shelf seas (Thomas and Schneider 1999; Bozec et al. 2006; Prowe et al. 2009). DIC concentrations in the Baltic Sea surface waters range from 18–21 mg dm^{-3} in the southwestern parts to 16–18 mg dm^{-3} in the northern Gulf of Bothnia (Thomas and Schneider 1999).

DOC is the second largest carbon reservoir in the Baltic. Its concentrations in the surface water vary from 3.0–4.3 mg dm^{-3} for the open waters of the Gulf of Bothnia and the southwestern parts of the Baltic to 3.5–6.5 mg dm^{-3} for the vicinity of estuaries. The concentrations are 3–4 times higher than those reported for the open ocean water (Voipio 1981; Pempkowiak et al. 1984; Ferrari et al. 1996; Hagström et al. 2001; Algesten et al. 2006; Kuliński and Pempkowiak 2008). POC exhibits the largest relative seasonal oscillations of concentrations. This is caused primarily by intensive phytoplankton bloom events, when the POC concentration exceeds 1.0 mg dm^{-3}. During the winter season, POC concentrations in the Baltic open water, that do not exceed 0.1 mg dm^{-3}, are controlled mainly by the amount of slowly sinking detritus (Pempkowiak et al. 1984; Burska et al. 2005; Dzierzbicka-Glowacka et al. 2010).

2.3.3 Carbon Sources and Losses in the Baltic Sea

As a result of processes occurring in the water column (e.g., photosynthesis, respiration, sorption), carbon species transform into one another without changing the total carbon concentration. There are additional external factors that also influence carbon reservoirs in the Baltic ecosystem compartments. Depending on the resulting flux direction, the factors are classified as sources—causing the concentration increase, and losses—reducing carbon concentrations within the basin. Overview of carbon major sources and sinks in the Baltic Sea is presented in the following sections.

2.3.3.1 Carbon Exchange Between the Baltic and the North Sea

The Danish Straits are the only route of water exchange between the Baltic Sea and the North Sea, and further with the Atlantic Ocean. Along with the sea water masses, dissolved and suspended substances, including carbon species, are transported. The amount of carbon that is exported from, and imported to, the Baltic depends mainly on the volume of flowing water. According to various sources, as reviewed and summarized by Omstedt et al. (2004), it has been demonstrated that, on average, 1,100–2,520 km^3 of water is yearly discharged from the Baltic to the North Sea, while 600–1,350 $km^3\ year^{-1}$ is transported in the opposite direction. These significant data differences result from a high variability of the direction and volume of flowing water masses, that are dependent on wind velocity and direction, and are often changing in a short time intervals (BACC Author Team 2008). Döös et al. (2004) estimate that only 6% of water flowing through the Great Belt resides in the Baltic for longer than a year, compared to 32% for the Oresund. Therefore, the BACC Author Team (2008) recommends the salinity to be used as an indicator allowing for water masses distinction and determination of the magnitude of flow through the Danish Straits. Despite their significant variability, the above-mentioned values indicate that the water exchange between the two basins is a decisive factor determining the Baltic Sea water balance. At the same time, the carbon exchange through the Danish Straits becomes an important aspect of the Baltic Sea carbon cycle. Based on calculations of water mass flows and their patterns (Wulff et al. 2001), coupled with an assumption of constant DIC and DOC annual concentrations in the Baltic and the North Sea water, Thomas et al. (2003) estimated that Baltic exports yearly 20 Tg C, in contrast to the yearly import amounting to just 6 Tg C. Based on these data, Thomas et al. (2005) suggested the Baltic Sea to be the largest net source of carbon for the North Sea. Moreover, both basins, combined, play the essential role in the mechanism of carbon transfer from the atmosphere and land to the deeper water of the North Atlantic.

2.3.3.2 Carbon Supplied With the River Runoff

Rivers are considered to be an essential element of the water balance of the Baltic Sea and important source of carbon. Pempkowiak and Kupryszewski (1980) estimated that, at least, 25% of organic matter present in the Baltic Sea has a terrestrial origin, out of which 50% is represented by humic substances, resistant to biochemical degradation (Pempkowiak and Pocklington 1983). Despite this, only few studies have focused on this group of compounds, in contrast to detailed investigations concerning nutrients (e.g., Petterson et al. 1997; Wulff et al. 2001; Granskog et al. 2005a).

River waters discharged into the Baltic are characterized by high concentrations of organic carbon (Granskog et al. 2005b). Pempkowiak (1985) has shown that DOC concentration in the lower stretch of the Vistula River ranges from 7.2 mg dm^{-3} in winter to 9.4 mg dm^{-3} in summer. Assuming the same DOC concentrations for all the remaining rivers flowing into the Baltic, it was estimated that rivers bring annually a total of 3.35 Tg of organic carbon, which represents approximately 3.3% of the total DOC pool in the Baltic. Similar DOC concentrations (9.5 mg dm^{-3}) were observed for the Svartan River and Öre River flowing into the Gulf of Bothnia (Langenheder et al. 2003; Granskog et al. 2005b). In a contrary, Granskog et al. (2005a) recorded DOC concentrations reaching as much as 14 mg dm^{-3} for the Siikajoki River, also flowing into the Gulf of Bothnia, which situates this river among watercourses characterized by the highest DOC concentrations in the world (Cauwet 2002; Dittmar and Kattner 2003). High concentrations of organic carbon in rivers flowing into the Gulf of Bothnia make them the source of a significant carbon load for the Baltic Sea. Algesten et al. (2006) estimated that the rivers escaping into the Baltic within Finland and Sweden deliver, respectively, 0.65 and 0.85 Tg of organic carbon annually.

In contrast to numerous reports on organic carbon, the data concerning inorganic carbon in the Baltic rivers are rather limited (Pempkowiak 1985; Thomas and Schneider 1999). Alkalinity is the parameter used in river monitoring programs, that allows to estimate DIC concentration in the most accurate manner. Alkalinity is a measure of the ability of a solution to neutralize hydrogen ions, expressed in equivalent mass per unit volume. In the surface water, it is determined mainly by the presence and concentration of CO_3^{2-} and HCO_3^- ions. In the rivers escaping to the Baltic, due to pH ranging from 7.0 to 7.5, the HCO_3^- ions play the dominant role.

The Scandinavian rivers are characterized by relatively low DIC concentrations, and at the same time low alkalinity. Brink et al. (2007) determined an average alkalinity for the Kalix and Lule rivers during 1985–2003 at 0.22 and 0.15 meq dm^{-3} respectively. When assuming that alkalinity is driven by the presence of HCO_3^- the value is equal to DIC concentrations corresponding to 2.6 and 1.8 mg dm^{-3} for Kalix and Lule, respectively. The DIC present in the rivers originates mostly from the organic matter mineralization, and to a smaller extent from weathering and erosion. The authors estimate that 1.9 and 1.2 mg DIC dm^{-3}, respectively, for the Kalix and the Lule, come from mineralization of organic matter.

Thomas and Schneider (1999) obtained similar DIC concentration values (approximately 2 mg dm^{-3}) for river discharges flowing into the Gulf of Bothnia. Relationship of DIC concentrations and salinity was determined for different parts of the open Baltic Sea water, and the correlations obtained for each zone were extrapolated to 0 PSU, giving DIC concentration in river water. For the Gulf of Finland and the Gulf of Riga, higher DIC concentrations have been estimated, amounting to 7 and 33 mg dm^{-3}, respectively. Water of the river mouth in the southern Baltic is also characterized by high DIC concentrations. Unpublished data (Beldowski, personal communication) indicate relatively constant DIC concentrations for the Świna (24–27 mg dm^{-3}) and well pronounced seasonal dynamics of DIC in the Vistula (32–45 mg dm^{-3}). Based on extrapolation method described by Thomas and Schneider (1999), and DOC average concentration of 4.3 mg dm^{-3}, it has been estimated (Thomas et al. 2003) that continental rivers supply the Baltic

Sea with 4.9 Tg C per annum. This value represents approximately 35-fold higher carbon load than that estimated for the Scandinavian rivers (0.12 Tg C year^{-1}; Thomas et al. 2003).

2.3.3.3 Carbon Deposition to the Sediments

POC plays a dominant role in carbon transport to the sea floor, while it is assumed that the role of PIC is minimal and falls within the error limits (Zsolnay 1973; Schneider et al. 2000; Struck et al. 2004). Carbon deposition depends on a number of processes taking place in a water column and sediments. The amount and quality of the material that can, potentially, sink to the bottom is the most significant factor. The main sources of organic matter containing particles suspended in the water column are the primary productivity and fluvial processes. As a result of mineralization and decomposition, only a small fraction of the suspended matter reaches the sediments, and the actual deposition location is determined by hydrological factors (Voipio 1981; Blomqvist and Heiskanen 2001).

The primary productivity magnitude in the Baltic changes spatially and temporarily. Nutrients are the driving force for primary productivity: the Gulf of Bothnia is the least active basin in this respect, while the greatest primary productivity is observed for the Gulf of Riga and the south-western part of the Baltic (Table 2.1;. Voipio 1981; Wasmund et al. 2001; HELCOM 2009).

The comparison of data on the primary productivity, as summarized by Wasmund and Siegel (2008), revealed the significant increase in the last decades in the southern parts of the Baltic and in the Gulf of Riga, while no significant changes have been reported for the Gulf of Bothnia (Table 2.1). The largest increase (nearly threefold) was observed in the Gulf of Riga, somewhat smaller (approximately twofold) in the Danish Straits and the so-called Baltic Proper. Changes in the magnitude of primary productivity for the entire basin lead to its overall increase from 84 g m^{-2} year^{-1} in the late 1970s and 1980s to 150 g m^{-2} year^{-1} presently.

Table 2.1 Primary productivity in the Baltic Sea (Wasmund and Siegel 2008 based on Kaiser et al. 1981 and Wasmund et al. 2001)

Region	Primary productivity (g C m^{-2} year^{-1})	
	Kaiser et al. (1981)	Wasmund et al. (2001)
Kattegat/Danish straits	90–120	190
Proper Baltic	90–125	200
Gulf of Riga	80–100	261
Gulf of Finland	70	82
Bothnian Sea	70	52
Gulf of Bothnia	18	17
Baltic Sea—average	84	150

Organic carbon is deposited mainly in areas of fine-grained sediments (mud-clay) covering 32% of the Baltic Sea bottom (Fig. 2.12; Emeis et al. 2000; Al-Hamdani and Reker 2007). This type of sediments is characterized by high organic carbon concentrations, reaching 10–11% in the uppermost sediment layers. Such high concentrations result from sedimentation of material that is rich in organic matter, and the permanent oxygen deficit in the sediments. The absence of oxygen slows down the mineralization rate of organic matter. The regions characterized by oxygen deficit and substantial sedimentation include: the Gotland Deep, the Gdansk Deep and the Bornholm Deep, where more than 50% of the Baltic total organic matter is deposited (Emeis et al. 2000). Areas of sea bottom

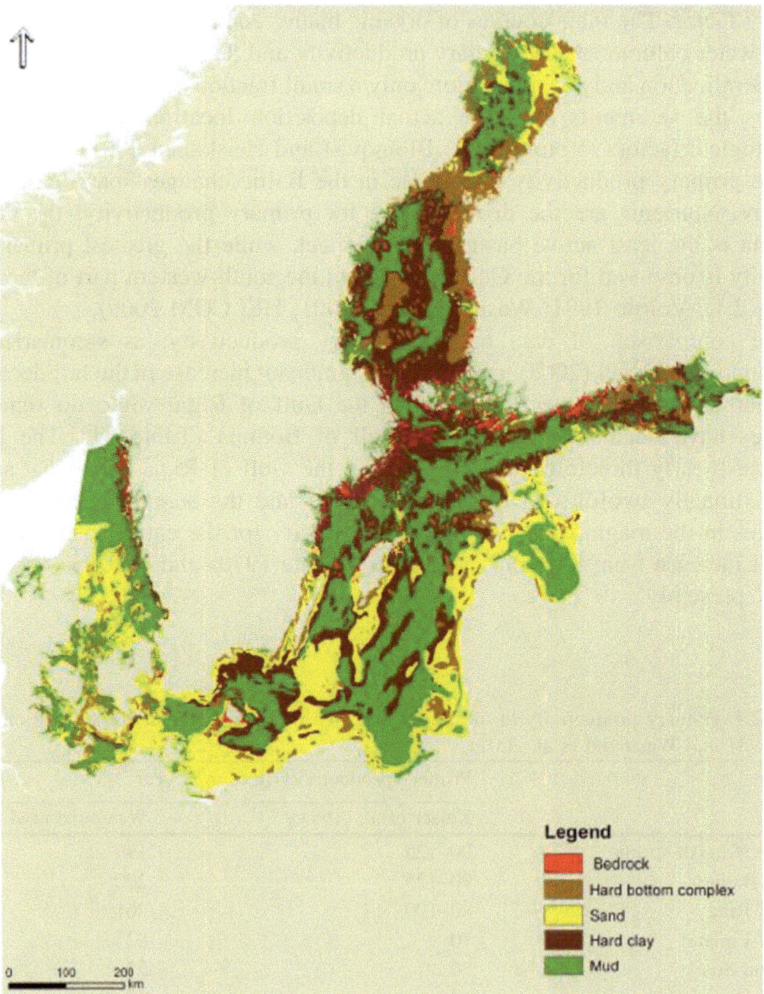

Fig. 2.12 The Baltic Sea sediment types (Al-Hamdani and Reker 2007)

covered by other sediment types play a minor role as accumulation areas of sedimentary material in the Baltic, and the majority of organic matter reaching their surface is subjected to rapid mineralization or resuspension and relocation to the depositional areas. This is best indicated by low concentrations of organic carbon within the uppermost sediment layers: 0.05–0.5% of dry weight (d.w.; Maksymowska 1998).

Studies involving sediment traps conducted in the Gotland Deep (Struck et al. 2004) revealed that POC transport to the sediments at a depth of 140 m amounts to 6.0 g m^{-2} $year^{-1}$. Summer and autumn carbon fluxes contribute most significantly to the annual POC flux: 31.9% and 34.6%, respectively. Slightly higher levels of POC transport to the sediments (also assessed using sediment traps) obtained by Schneider et al. (2000) were calculated to be 7.9 g m^{-2} $year^{-1}$. Based on this value, assumed to be the same for the Baltic Proper, the Gulf of Finland and the Gulf of Riga, Schneider et al. (2000) estimated that 1.7 Tg C is deposited in these regions every year.

The total organic carbon (TOC) accumulation rate in the bottom sediments is calculated based on the designated sedimentation rate, usually assessed using a method based on ^{210}Pb activity, as well as organic carbon concentration levels in the sediments. Algesten et al. (2006) estimated, based on the accumulation rate of TOC for the uppermost sediment layers, that 1.1 Tg C is deposited in the Gulf of Bothnia annually. On the other hand, Emeis et al. (2000) indicate that TOC accumulation rates in the Bornholm Deep, Gdansk Deep, and Gotland Deep increased two- to three-fold during the several last decades, while current values amount to 60 g C m^{-2} $year^{-1}$, 60–65 g C m^{-2} $year^{-1}$ and 40 g C m^{-2} $year^{-1}$, respectively. Only in the vicinity of Arkona Deep, organic carbon accumulation rates remain at a constant level ranging from 20 to 25 g C m^{-2} $year^{-1}$. The suggested increase of TOC accumulation rates is attributed to a greater organic matter influx to the sediments and transformations of environmental conditions determining organic matter degradation processes (Emeis et al. 2000; Struck et al. 2000, Szczepanska et al. 2011). This opinion is also confirmed by Jonsson and Carman's (1994) studies that estimated close to twofold increase of TOC accumulation rates in the Gulf of Bothnia in the 1980s, compared to the 1920s. For the Baltic Proper, this increase has been estimated to be five to ten times higher than the historical one.

2.3.3.4 Carbon Dioxide Exchange at the Water/Atmosphere Interface

The Baltic Sea is a diverse ecosystem. The basin consists of regions characterized by a low primary productivity and large terrestrial organic matter inflow (the Gulf of Bothnia), through highly productive areas, located in the central and southern part of the basin, to highly eutrophic regions specific for the main rivers mouths (Voipio 1981; Hagström et al. 2001; Nausch et al. 2008; Wasmund and Siegel 2008). This situation is reflected in the direction and rate of CO_2 exchange through the water/atmosphere interface.

The Gulf of Bothnia is considered to be a source of CO_2 to the atmosphere, which is attributed to intensive mineralization of organic matter supplied to the Gulf with the river runoff and limited, especially when compared with the southern Baltic, phytoplankton activity (Algesten et al. 2004 and Algesten et al. 2006). It has been estimated (Algesten et al. 2004) that the Gulf of Bothnia annually releases 3.79 Tg C in the form of CO_2, which corresponds to 37.2 g C m^{-2}. Subsequent estimates concerning CO_2 cycling in the Gulf of Bothnia were similar. Algesten et al. (2006) calculated carbon emission for the entire Gulf at the level of 3.61 Tg C $year^{-1}$, which corresponds to specific emissions of 35.4 g C m^{-2} $year^{-1}$.

In contrast to the Gulf of Bothnia, the remaining portion of the Baltic Sea is considered to absorb atmospheric CO_2 (Ohlson 1990; Thomas and Schneider 1999; Kuss et al. 2006). Ohlson (1990) estimated the average atmospheric CO_2 uptake at 13.2 g C m^{-2} $year^{-1}$. Thomas and Schneider (1999) obtained comparable results, and based on biogeochemical model the authors calculated that the so-called Baltic Proper (Baltic without the major gulfs: Bothnia, Finland, Riga and Danish Straits) absorbs 10.8 $\pm$ 1.1 g C m^{-2} $year^{-1}$. In the southern Baltic, the Arkona Deep is of special interest; buoy-based measurements of partial pressure of dissolved CO_2 in water are available for this location (Kuss et al. 2006). The results indicate that this region is characterized by carbon dioxide absorption amounting to 36.0 g C m^{-2} $year^{-1}$.

2.3.3.5 Other Sources and Losses of Carbon in the Baltic Sea

Additional sources of carbon in the Baltic Sea include precipitation and the so-called point sources (HELCOM 2004; Algesten et al. 2006). These consist of all terrestrial carbon inflows, other than river runoff. Wastewater treatment plants located along the shoreline are the main source (70%) here (HELCOM 2004). The data indicate that annually 3.6 km^3 of treated sewage is discharged to the Baltic. Organic carbon load delivered from the point sources is frequently expressed as the seven-day biochemical oxygen demand (BOD_7). This, for the entire Baltic Sea, is calculated to be 88.7 Gg O_2. Converted to the carbon load, using a BOD_7/C_{org} ratio of 2.28 (HELCOM 1983), this corresponds to 39 Gg C, half of which is delivered to the Gulf of Bothnia. Industrial plants, associated mainly with paper and timber industry, contribute 65% of the total organic carbon inflow from the point sources. To a smaller extent, supply from point sources includes municipal wastewater (33%) and fish farms (2%).

Precipitation constitutes an important freshwater source for the Baltic Sea. However, the literature data lack concentrations of organic and inorganic carbon in the rainwater within the Baltic. Based on the data presented by Anttila et al. (1995) and Algesten et al. (2006) the estimated amount of organic carbon reaching the Gulf of Bothnia with rainwater at 0.11 Tg $year^{-1}$.

On the other hand, fisheries cause carbon depletion in the Baltic. Since fish tissue comprise significant amounts of organic carbon, it can be regarded as a reservoir of carbon. Herring (*Clupea harengus membras*) and sprat (*Sprattus sprattus*) are the most commonly harvested species—presently contributing over 80% of the total catch. Fish harvested to a smaller extent include cod, flounder, salmon and freshwater species (Hammer et al. 2008; ICES 2008). According to the ICES report (ICES 2008), the catch magnitude in the period 1974–1984 oscillated at a level of 850–990 10^3 ton $year^{-1}$, increasing during 1996–1998, primarily due to cod catch, to 1,100 10^3 tons $year^{-1}$. ICES (2008) estimates the current Baltic Sea annual catch to be 700 10^3 ton fish $year^{-1}$. Assuming 8.2% as an average carbon concentration in a wet fish tissue (Crabtree 1995), this corresponds to carbon extraction from the Baltic equal to 57 Gg C $year^{-1}$.

References

Algesten G, Wikner J, Sobek S, Tranvik LJ, Jansson M (2004) Seasonal variation of CO_2 saturation in the Gulf of Bothnia: indications of marine net heterotrophy. Glob Biogeochem Cycles 18:GB4021

Algesten G, Brydsten L, Jonsson P, Kortelainen P, Löfgren S, Rahm L, Räike A, Sobek S, Tranvik L, Wikner J, Jansson M (2006) Organic carbon budget for the Gulf of Bothnia. J Mar Syst 63:155–161

Al-Hamdani Z, Reker J (eds) (2007) Towards marine landscapes in the Baltic Sea. BALANCE interim report 10, Geological Survey of Denmark and Greenland, Kopenhagen, p 118

Alley RB, Clark PU, Huybrechts P, Joughin I (2005) Ice-sheet and sea-level changes. Science 310:456–460

Andersson AJ, Mackenzie FT (2004) Shallow-water oceans: a source or sink of atmospheric CO_2? Front Ecol Environ 2:348–353

Andersson A, Rudehäll Å (1993) Proportion of plankton biomass in particulate organic carbon in the northern Baltic Sea. Mar Ecol Prog Ser 95:133–139

Andreae OA (2007) Athmosphere: aerosols before pollution. Science 315:50–51

Andreae MO, Merlet P (2001) Emission of trace gases and aerosols from biomass burning. Glob Biogeochem Cycles 15:955–966

Anisimov OA, Vaughan DG, Callaghan TV, Furgal C, Marchant H, Prowse TD, Vilhjálmsson H, Walsh JE (2007) Polar regions (Arctic and Antarctic). In: Parry ML, Canziani OF, Palutikof JP, van der Linden PJ, Hanson CE (eds) Climate change: impacts, adaptation and vulnerability: contribution of working group II to the fourth assessment report of the intergovernmental panel on climate change. Cambridge University Press, Cambridge, pp 653–685

Anttila P, Paatero P, Tapper U, Järvinen O (1995) Source identification of bulk wet deposition in Finland by positive matrix factorization. Atmos Environ 29:1705–1718

BACC Author Team (2008) Assessment of climate change for the Baltic Sea basin. Springer, Berlin, p 473

Bański J (2009) Zmiany gospodarki rolnej i zabezpieczenia żywnościowe ludności w wyniku zmian klimatycznych (Changes in agricultural economy and food supply resulting from climate changes). In: Gutry-Korycka M, Markowski T (eds) Zrównoważone warunki życia w zmieniającym się systemie klimatycznym Ziemi. Studia Tom CXXIV. Komitet Przestrzennego Zagospodarowania Kraju PAN, Warszawa, pp 64–74 (in Polish)

Bates J (1987) Milankovitch and climate. Earth Sci Rev 24:137–139

Bates NR, Merlivat L, Beaumont L, Pequignet AC (2000) Intercomparison of shipboard and moored CARIOCA buoy seawater fCO_2 measurements in the Sargasso Sea. Mar Chem 72:239–255

Beniston M (2007) Linking extreme climate events and economic impacts: examples from the Swiss Alps. Energy Policy 35:5384–5392

Benner R (2002) Chemical composition and reactivity. In: Hansell DA, Carlson CA (eds) Biogeochemistry of marine dissolved organic matter. Elsevier Science, San Diego, 59–90
Bergström S, Alexandersson H, Carlsson B, Josefsson W, Karlsson K-G, Westring G (2001) Climate and hydrology of the Baltic basin. In: Wulff FV, Rahm LA, Larsson P (eds) A system analysis of the Baltic Sea. Springer, Berlin, pp 75–112
Bindoff NL, Willebrand J, Artale V, Cazenave A, Gregory J, Gulev S, Hanawa K, Le Quéré C, Levitus S, Nojiri Y, Shum CK, Talley LD, Unnikrishnan A (2007) Observations: oceanic climate change and sea level. In: Solomon S, Qin D, Manning M, Chen Z, Marquis M, Averyt KB, Tignor M, Miller HL (eds) Climate change: the physical science basis: contribution of working group I to the fourth assessment report of the intergovernmental panel on climate change. Cambridge University Press, Cambridge, pp 385–432
Błażejczyk K (2009) Zmiany globalne klimatu i ich konsekwencje zdrowotne dla człowieka (Global climate changes and their consequences for human health). In: Gutry-Korycka M, Markowski T (eds) Zrównoważone warunki życia w zmieniającym się systemie klimatycznym Ziemi. Studia Tom CXXIV. Komitet Przestrzennego Zagospodarowania Kraju PAN, Warszawa, pp 107–136 (in Polish)
Blomqvist S, Heiskanen A-S (2001) The challenge of sedimentation in the Baltic Sea. In: Wulff FV, Rahm LA, Larsson P (eds) A system analysis of the Baltic Sea. Springer, Berlin, pp 211–227
Borges AV (2005) Do we have enough pieces of the jigsaw to integrate CO_2 fluxes in the coastal ocean? Estuaries 28:3–27
Borges AV, Delille B, Frankignoulle M (2005) Budgeting sinks and sources of CO_2 in the coastal ocean: diversity of ecosystems counts. Geophys Res Lett 32:L14601
Borges AV, Schiettecatte L-S, Abril G, Delille B, Gazeau F (2006) Carbon dioxide in European coastal waters. Estuar Coast Shelf Sci 70:375–387
Bozec Y, Thomas H, Schiettecatte L-S, Borges AV, Elkalay K, de Baar HJW (2006) Assessment of the processes controlling seasonal variations of dissolved inorganic carbon in the North Sea. Limnol Oceanogr 51:2746–2762
Breiling M, Charamza P (1999) The impact of global warming on winter tourism and skiing: a regionalised model for Austrian snow conditions. Reg Environ Change 1:4–14
Brink J, Humborg C, Sahlberg J, Rahm L, Mörth C-M (2007) Weathering rates and origin of inorganic carbon as influenced by river regulation in the boreal sub-arctic region of Sweden. Hydrol Earth Syst Sci Discuss 4:555–588
Burska D, Pryputniewicz D, Falkowska L (2005) Stratification of particulate organic carbon and nitrogen in the Gdańsk Deep (southern Baltic Sea). Oceanologia 47:201–217
Cai W-J, Dai M (2004) Comment on "Enhanced open ocean storage of CO_2 from shelf area pumping" by Thomas et al (Science 304). Science 306:1477
Cai WJ, Dai MH, Wang YC (2006) Air-sea exchange of carbon dioxide in ocean margins: a province-based synthesis. Geophys Res Lett 33:L12603
Caldeira K, Wickett ME (2003) Anthropogenic carbon and ocean pH. Nature 425:365
Carlson CA (2002) Production and removal processes. In: Hansell DA, Carlson CA (eds) Biogeochemistry of marine dissolved organic matter. Elsevier Science, San Diego, pp 91–151
Cauwet G (2002) DOM in the coastal zone. In: Hansell DA, Carlson CA (eds) Biogeochemistry of marine dissolved organic matter. Elsevier Science, San Diego, pp 579–609
Charlson RJ, Schwartz SE, Hales JM, Cess RD, Coakley JA Jr, Hansen JE, Hofmann DJ (1992) Climate forcing by anthropogenic aerosols. Science 24:423–430
Chen C-TA, Borges AV (2009) Reconciling opposing views on carbon cycling in the coastal ocean: continental shelves as sinks and near-shore ecosystems as sources of atmospheric CO_2. Deep-Sea Res II 56:578–590
Chen C-TA, Liu K-K, Macdonald R (2003) Continental margin exchanges. In: Fasham MJR (ed) Ocean biogeochemistry. Springer, Berlin, pp 53–97
Chisholm SW (2000) Stirring times in the southern ocean. Nature 407:685–686
Clark PU, Mix AC (2000) Ice sheets by volume. Nature 406:689–690
Clark PU, Alley RB, Pollard D (1999) Northern hemisphere ice sheet influences on global climate change. Science 286:1103–1111

Colijn F, Cadée GC (2003) Is phytoplankton growth in the Wadden Sea light or nitrogen limited? J Sea Res 49:83–93

Confalonieri U, Menne B, Akhtar R, Ebi KL, Hauengue M, Kovats RS, Revich B, Woodward A (2007) Human health. In: Parry ML, Canziani OF, Palutikof JP, van der Linden PJ, Hanson CE (eds) Climate change: impacts, adaptation and vulnerability: contribution of working group II to the fourth assessment report of the intergovernmental panel on climate change. Cambridge University Press, Cambridge, pp 391–431

Crabtree RE (1995) Chemical composition and energy content of deep-sea demersal fishes from tropical and temperate regions of the western North Atlantic. Bull Mar Sci 56:434–449

de Haas H, van Weering TCE, de Stigter H (2002) Organic carbon in shelf seas: sinks or sources, processes and products. Cont Shelf Res 22:691–717

Dittmar T, Kattner G (2003) The biogeochemistry of the river and shelf ecosystem of the Arctic Ocean: a review. Mar Chem 83:103–120

Dlugokencky EJ, Houweling S, Bruhwiler L, Masarie KA, Lang PM, Miller JB, Tans PP (2003) Atmospheric methane levels off: temporary pause or a new steady-state? Geophys Res Lett 30:GL018126

Doney SC (2006) Oceanography: plankton in a warmer world. Nature 444:695–696

Doney SC, Lindsay K, Moore JK (2003) Global ocean carbon cycle modeling. In: Fasham MJR (ed) Ocean biogeochemistry. Springer, Berlin, pp 217–238

Döös K, Meier HEM, Döscher R (2004) The Baltic haline conveyor belt or the overturning circulation and mixing in the Baltic. Ambio 33:261–266

Dzierzbicka-Głowacka L, Kuliński K, Maciejewska A, Jakacki J, Pempkowiak J (2010) Particulate organic carbon in the southern Baltic Sea: numerical simulations and experimental data. Oceanologia 52(4):621–648

Easterling WE, Aggarwal PK, Batima P, Brander KM, Erda L, Howden SM, Kirilenko A, Morton J, Soussana J-F, Schmidhuber J, Tubiello FN (2007) Food, fibre and forest products. In: Parry ML, Canziani OF, Palutikof JP, van der Linden PJ, Hanson CE (eds) Climate change: impacts, adaptation and vulnerability: contribution of working group II to the fourth assessment report of the intergovernmental panel on climate change. Cambridge University Press, Cambridge, pp 273–313

Emeis K-C, Struck U, Leipe T, Pollehne F, Kunzendorf H, Christiansen C (2000) Changes in the C, N, P burial rates in some Baltic Sea sediments over the last 150 years—relevance to P regeneration rates and the phosphorus cycle. Mar Geol 167:43–59

Emerson SR, Hedges JI (2008) Chemical oceanography and the marine carbon cycle. Cambridge University Press, Cambridge, p 453

Fairbanks RG (1989) A 17, 000-year glacio-eustatic sea level record: influence of glacial melting rates on the Younger Dryas event and deep-ocean circulation. Nature 342:637–642

Feely RA, Sabine CL, Lee K, Berelson W, Kleypas J, Fabry VJ, Millero FJ (2004) Impact of anthropogenic CO_2 on the $CaCO_3$ system in the oceans. Science 305:362–366

Fenger J, Buch E, Jacobsen PR (2001) Monitoring and impacts of sea level rise at Danish coasts and near shore infrastructures. In: Jorgensen AM, Fenger J, Halsnes K (eds) Climate change research—Danish contributions. Danish Climate Centre, Copenhagen, pp 237–254

Ferrari GM, Dowell MD, Grossi S, Targa C (1996) Relationship between the optical properties of chromophoric dissolved organic matter and total concentration of dissolved organic carbon in the southern Baltic Sea region. Mar Chem 55:299–316

Fischlin A, Midgley GF, Price JT, Leemans R, Gopal B, Turley C, Rounsevell MDA, Dube OP, Tarazona J, Velichko AA (2007) Ecosystems, their properties, goods, and services. In: Parry ML, Canziani OF, Palutikof JP, van der Linden PJ, Hanson CE (eds) Climate change: impacts, adaptation and vulnerability: contribution of working group II to the fourth assessment report of the intergovernmental panel on climate change. Cambridge University Press, Cambridge, pp 211–272

Forster P, Ramaswamy V, Artaxo P, Berntsen T, Betts R, Fahey DW, Haywood J, Lean J, Lowe DC, Myhre G, Nganga J, Prinn R, Raga G, Schulz M, Van Dorland R (2007) Changes in atmospheric constituents and in radiative forcing. In: Solomon S, Qin D,

Manning M, Chen Z, Marquis M, Averyt KB, Tignor M, Miller HL (eds) Climate change: the physical science basis: contribution of working group I to the fourth assessment report of the intergovernmental panel on climate change. Cambridge University Press, Cambridge, pp 129–234

Frudenthal T, Wagner T, Wenzhöfer F, Zabel M, Wefer G (2001) Early diagenesis of organic matter from sediments of the eastern subtropical Atlantic: evidence from stable nitrogen and carbon isotopes. Geochim Cosmochim Acta 65:1795–1808

Ganopolski A, Rahmstorf S (2001) Simulation of rapid glacial climate changes in a coupled climate model. Nature 409:153–158

Gao C, Robock A, Ammann C (2008) Volcanic forcing of climate over the past 1,500 years: an improved ice core-based index for climate models. J Geophys Res 113:D23111

Gardner WD, Mishonova AV, Richardson MJ (2006) Global POC concentrations from in situ and satellite data. Deep-Sea Res II 53:718–740

Gattuso J-P, Frankignoulle M, Wollast R (1998) Carbon and carbonate metabolism in coastal aquatic ecosystems. Annu Rev Ecol Syst 29:405–434

Georges C (2004) The 20th century glacier fluctuations in the Cordillera Blanca (Perú). Arct Antarct Alp Res 36:100–107

Godoi RHM, Aerts K, Harlay J, Kaegi R, Ro C-U, Chou L, Van Grieken R (2009) Organic surface coating on Coccolithophores—*Emiliania huxleyi*: its determination and implication in the marine carbon cycle. Microchem J 91:266–271

Gordon JE, Haynes VM, Hubbard A (2008) Recent glacier changes and climate trends on South Georgia. Glob Plan Change 60:72–84

Granskog MA, Ehn J, Niemalä M (2005a) Characteristics and potential impacts of under-ice river plumes in the seasonally ice-covered Bothnian Bay (Baltic Sea). J Mar Syst 53:187–196

Granskog MA, Kaartokallio H, Thomas DN, Kuosa H (2005b) Influence of freshwater inflow on the inorganic nutrient and dissolved organic matter within coastal sea ice and underlying waters in the Gulf of Finland (Baltic Sea). Estuar Coast Shelf Sci 65:109–122

Greatbatch RJ (2000) The North Atlantic oscillation. Stoch Env Res Risk Assess 14:213–242

Groisman PY, Karl TR, Knight RW (1994) Observed impact of snow cover on the heat balance and the rise of continental spring temperatures. Science 263:198–200

Gudelis WK, Jemielianow JM (1982) Geologia Morza Bałtyckiego. Wydawnictwa Geologiczne, Warszawa, p 412

Gutry-Korycka M (2009) Zobowiązania nauki wobec zrównoważonego rozwoju środowiska życia człowieka w warunkach globalnych zmian klimatu (Science obligations concerning the sustainable development of human environment under the conditions of global climate change). In: Gutry-Korycka M, Markowski T (eds) Zrównoważone warunki życia w zmieniającym się systemie klimatycznym Ziemi. Studia Tom CXXIV. Komitet Przestrzennego Zagospodarowania Kraju PAN, Warszawa, pp 8–21 (in Polish)

Hagen JO, Melvold K, Pinglot F, Dowdeswell JA (2003) On the net mass balance of the glaciers and ice caps in Svalbard, Norwegian Arctic. Arct Antarct Alp Res 35:264–270

Hagström Å, Azam F, Kuparinen J, Zweifel U-L (2001) Pelagic plankton growth and resource limitations in the Baltic Sea. In: Wulff FV, Rahm LA, Larsson P (eds) A system analysis of the Baltic Sea. Springer, Berlin, pp 177–210

Hammer C, von Dorrien C, Ernst P, Gröhsler T, Köster F, Mackenzie B, Möllmann C, Wegner G, Zimmermann C (2008) Fish stock development under hydrographic and hydrochemical aspects, the history of the baltic sea fisheries and its management. In: Feistel R, Nausch G, Wasmund N (eds) State and evolution of the Baltic Sea, 1952–2005. Wiley, Hoboken, pp 543–581

Hansell DA (2002) DOC in the Global Ocean carbon cycle. In: Hansell DA, Carlson CA (eds) Biogeochemistry of marine dissolved organic matter. Elsevier Science, San Diego, pp 685–715

Hansell DA, Carlson CA (1998) Net community production of dissolved organic carbon. Glob Biogeochem Cycles 12:443–453

Hansen J, Nazarenko L (2004) Soot climate forcing via snow and ice albedos. Proc Natl Acad Sci U S A 101:423–428

Harff J, Frischbutter A, Lampe R, Meyer M (2001) Sea level changes in the Baltic Sea–interrelation of climatic and geological processes. In: Gerhard LC, Harrison WE, Hanson BM (eds) Geological perspectives of global climate change (AAPG-Studies in Geology, vol 47), pp 231–250

Hedges JI (2002) Why dissolved organic matter? In: Hansell DA, Carlson CA (eds) Biogeochemistry of marine dissolved organic matter. Elsevier Science, San Diego, pp 1–33

Hedges JI, Keil R, Benner R (1997) What happens to terrestrial organic matter in the ocean? Org Geochem 27:195–212

HELCOM (1983) Seminar on review of progress made in water protection measures. In: Baltic Sea environment proceedings, vol 14, p 436

HELCOM (2002) Environment of the Baltic Sea area 1994–1998. In: Baltic Sea environment proceedings, vol 82B, p 214

HELCOM (2004) The fourth Baltic Sea pollution load compilation (PLC-4). In: Baltic Sea environment proceedings, vol 93, p 189

HELCOM (2007) Climate change in the Baltic Sea area. In: Baltic Sea environment proceedings, vol 111, p 49

HELCOM (2009) Eutrophication in the Baltic Sea. In: Baltic Sea environment proceedings, vol 115B, p 150

Hensen C, Zabel M, Schulz HN (2006) Benthic cycling of oxygen, nitrogen and phosphorus. In: Schulz HD, Zabel M (eds) Marine geochemistry. Springer, Berlin, pp 207–240

Hiscock WT, Millero FJ (2002) Nutrient and carbon parameters during the southern ocean iron experiment (SOFeX). Deep-Sea Res I 52:2086–2108

Honjo S, Manganini SJ, Krishfield RA, Francois R (2008) Particulate organic carbon fluxes to the ocean interior and factors controlling the biological pump: a synthesis of global sediment trap programs since 1983. Prog Oceanogr 76:217–285

Houghton RA (2003) Revised estimates of the annual net flux of carbon to the atmosphere from changes in land use and land management. Tellus 55B:378–390

Houghton RA (2007) Balancing the global carbon budget. Annu Rev Earth Plan Sci 35:313–347

Huertas IE, Ríos AF, García-Lafuente J, Makaoui A, Rodríguez-Gálvez S, Sánchez-Román A, Orbi A, Ruíz J, Pérez FF (2009) Anthropogenic and natural CO2 exchange through the strait of Gibraltar. Biogeosciences 6:647–662

Hurrell JW (1995) Decadal trends in the North Atlantic oscillation: regional temperatures and precipitation. Science 269:676–679

ICES (2008) Report of the ICES Advisory Committee, ICES Advice, Book 8, p 133

IGAC (2006) Science plan and implementation strategy. IGBP report number 56. IGBP Secretariat, Sztokholm, p 44

IGBP (2002) The Amsterdam IGBP declaration (June 2001). Pap Glob Change IGBP 9:9–10

IPCC (2001) Climate change: synthesis report: a contribution of working groups I, II and III to the third assessment report of the intergovernmental panel on climate change. Cambridge University Press, Cambridge, p 398

IPCC (2007) Climate change: synthesis report: a contribution of working groups I, II and III to the fourth assessment report of the intergovernmental panel on climate change. Cambridge University Press, Cambridge, p 73

Ito A, Penner JE (2005) Historical emissions of carbonaceous aerosols from biomass and fossil fuel burning for the period 1870–2000. Glob Biogeochem Cycles 19:GB2028

Jacobson MZ (2004) Climate response of fossil fuel and biofuel soot, accounting for soot's feedback to snow and sea ice albedo and emissivity. J Geophys Res 109:D21201

Johansson MM, Kahma KK, Boman H, Launiainen J (2004) Scenarios for Sea Level Rise on the Finnish Coast. Boreal Environ Res 9:153–166

Jonsson P, Carman R (1994) Changes in deposition of organic matter and nutrients in the Baltic Sea during the twentieth century. Mar Poll Bull 28:417–426

Kaiser W, Renk H, Schulz S (1981) Die Primärproduktion der Ostsee. Geodätische und Geophysikalische Veröffentlichungen. Reihe 4:27–52

Kalkstein LS, Barthel CD, Greene JS, Nichols MC (1996) A new spatial synoptic classification: application to air mass analysis. Int J Climatol 16:983–1004

Karl DM, Dore JE, Lukas R, Michaels AF, Bates NR, Knap A (2001) Building the long-term picture: the US JGOFS time-series programs. Oceanography 14:6–17

Keeling CD, Whorf TP, Wahlen M, van der Plicht J (1995) Interannual extremes in the rate of rise of atmospheric carbon dioxide since 1980. Nature 375:666–670

Kelley JJ, Hood DW (1971) Carbon dioxide in the Pacific Ocean and Bering Sea: upwelling and mixing. J Geophys Res 76:745–753

Key RM, Kozar A, Sabine CL (2004) A global ocean carbon climatology: results from global data analysis project (GLODAP). Glob Biogeochem Cycles 18:GB4031

Kiehl J, Trenberth K (1997) Earth's annual global mean energy budget. Bull Am Meteorol Soc 78:197–206

Kintisch E (2009) Projections of climate change go from bad to worse, scientists report. Science 323:1546–1547

Klein Goldewijk K (2001) Estimating global land use change over the past 300 years: the HYDE database. Glob Biogeochem Cycles 15:417–433

Klein AG, Kincaid JL (2006) Retreat of glaciers on Puncak Jaya, Irian Jaya, determined from 2000 and 2002 IKONOS satellite images. J Glaciol 52:65–79

Kożuchowski K (1998) Atmosfera, klimat, ekoklimat (Atmosphere, climate, ecoclimate). PWN, Warszawa, p 243 (in Polish)

Kuliński K, Pempkowiak J (2008) Dissolved organic carbon in the southern Baltic Sea: quantification of factors affecting its distribution. Estuar Coast Shelf Sci 78:38–44

Kuliński K, She J, Pempkowiak J (2011) Short and medium term dynamics of the carbon exchange between the Baltic Sea and the North Sea. Cont Schelf Res 31:1611–1619

Kundzewicz ZW (2009) Klęski żywiołowe—obserwacje, projekcje i możliwości przeciwdziałania (Natural disasters—observations, projections and preventive means). In: Gutry-Korycka M, Markowski T (eds) Zrównoważone warunki życia w zmieniającym się systemie klimatycznym Ziemi. Studia Tom CXXIV. Komitet Przestrzennego Zagospodarowania Kraju PAN, Warszawa, pp 22–39 (in Polish)

Kundzewicz ZW, Graczyk D, Pińskwar I, Radziejewski M, Szwed M, Bärring L, Giannakopoulos C, Holt T, Palutikof J, Leckerbusch GC, Ulbrich U, Schwarb M (2004) Changes in the occurrence of extremes part I: climatic background. Pap Glob Change IGBP 11:9–20

Kundzewicz ZW, Mata LJ, Arnell NW, Döll P, Kabat P, Jiménez B, Miller KA, Oki T, Sen Z, Shiklomanov IA (2007) Freshwater resources and their management. In: Parry ML, Canziani OF, Palutikof JP, van der Linden PJ, Hanson CE (eds) Climate change: Impacts adaptation and vulnerability: contribution of working group II to the fourth assessment report of the intergovernmental panel on climate change. Cambridge University Press, Cambridge, pp 173–210

Kuss J, Roeder W, Wlost KP, DeGrandpre MD (2006) Time-series of surface water CO_2 and oxygen measurements on a platform in the central Arkona Sea (Baltic Sea): seasonality of uptake and release. Mar Chem 101:220–232

Langenheder S, Kisand V, Wikner J, Tranvik LJ (2003) Salinity as a structuring factor for the composition and performance of bacterioplankton degrading riverine DOC. FEMS Microbiol Ecol 45:189–202

Lass H-U, Matthäus W (2008) General oceanography of the Baltic Sea. In: Feistel R, Nausch G, Wasmund N (eds) State and evolution of the Baltic Sea, 1952–2005. Wiley, Hoboken, pp 5–43

Le Treut H, Somerville R, Cubasch U, Ding Y, Mauritzen C, Mokssit A, Peterson T, Prather M (2007) Historical overview of climate change. In: Solomon S, Qin D, Manning M, Chen Z, Marquis M, Averyt KB, Tignor M, Miller HL (eds) Climate change: the physical science basis: contribution of working group I to the fourth assessment report of the intergovernmental panel on climate change. Cambridge University Press, Cambridge, pp 93–128

Lécuyer C, Simon L, Guyot F (2000) Comparison of carbon, nitrogen and water budgets on Venus and the Earth. Earth Plan Sci Lett 181:33–40

Lemke P, Ren J, Alley RB, Allison I, Carrasco J, Flato G, Fujii Y, Kaser G, Mote P, Thomas RH, Zhang T (2007) Observations: changes in snow, ice and frozen ground. In: Solomon S, Qin D, Manning M, Chen Z, Marquis M, Averyt KB, Tignor M, Miller HL (eds) Climate change: the physical science basis: contribution of working group I to the fourth assessment report of the intergovernmental panel on climate change. Cambridge University Press, Cambridge, pp 337–384

Liszewska M (ed) (2004) Potential climate changes and sustainable water management: publications of the Institute of Geophysics, vol E-4 (377). Polish Academy of Sciences, Warszawa, p 96

Lønborg C, Davidson K, Álvarez–Salgado XA, Miller AEJ (2009) Bioavailability and bacterial degradation rates of dissolved organic matter in a temperate coastal area during an annual cycle. Mar Chem 113:219–226

Longhurst AR, Harrison WG (1989) The biological pump: profiles of plankton production and consumption in the upper ocean. Prog Oceanogr 22:47–123

Lythe MB, Vaughan DG, BEDMAP Group (2001) BEDMAP: a new ice thickness and subglacial topographic model of Antarctica. J Geophys Res 106:11335–11351

MacKenzie BR, Gislason H, Möllmann C, Köster FW (2007) Impact of 21st century climate change on the Baltic Sea fish community and fisheries. Glob Change Biol 13:1348–1367

Maier-Reimer E, Hasselmann K (1987) Transport and storage of CO_2 in the ocean-an inorganic ocean-circulation carbon cycle model. Clim Dynam 2:63–90

Majewski W (2003) Powódź w Gdańsku w lipcu 2001 (Flood in Gdańsk in 2001). In: Cyberski J (ed) Powódź w Gdańsku 2001. Gdańskie Towarzystwo Naukowe, Gdańsk, pp 81–88 (in Polish)

Maksymowska D (1998) Degradacja materii organicznej w toni wodnej i osadach dennych Zatoki Gdańskiej (Degradation of organic matter in water and bottom sediments of the Bay of Gdańsk). PhD thesis, University of Gdańsk, p 147 (in Polish)

Martyn D (2000) Klimaty kuli ziemskiej (Climates of the Earth). PWN, Warszawa, p 359 (in Polish)

Matthäus W, Nehring D, Feistel R, Nausch G, Mohrholz V, Lass H-U (2008) The inflow of highly saline water into the Baltic Sea. In: Feistel R, Nausch G, Wasmund N (eds) State and evolution of the Baltic Sea, 1952–2005. Wiley, Hoboken, pp 265–309

McPhaden MJ, Zebiak SE, Glantz MH (2006) ENSO as an integrating concept in earth science. Science 314:1740–1745

Meier HEM (2007) Modeling the pathways and ages of inflowing salt- and freshwater in the Baltic Sea. Estuar Coast Shelf Sci 74:610–627

Meier HEM, Broman B, Kallio H, Klellstrom E (2004) Projections of future surface winds, sea levels and wind waves in the late 21st century and their application for impact studies of flood prone areas in the Baltic Sea region. In: Schmidt-Thomé P (ed) Special paper number 41, Geological Survey of Finland, Helsinki pp 88–101

Miętus M (1999) Rola integralnej cyrkulacji atmosferycznej w kształtowaniu warunków klimatycznych i oceanograficznych w polskiej strefie brzegowej Morza Bałtyckiego (Role of integral atmospheric circulation in shaping of climatic and oceanographic conditions in the Polish coastal zone of the Baltic Sea), IMGW, Warszawa, p 126 (in Polish)

Miętus M, Filipiak J, Owczarek M (2004) Klimat wybrzeża południowego Bałtyku. Stan obecny i perspektywy zmian (Climate of the southern Baltic coastal region: present state and expected changes). In: Cyberski J (ed) Środowisko Polskiej Strefy Południowego Bałtyku—stan obecny i przewidywane zmiany w przededniu integracji europejskiej. Wydawnictwo Gdańskie, Gdańsk, pp 11–44 (in Polish)

Miller L, Douglas BC (2004) Mass and volume contributions to 20th century global sea level rise. Nature 428:406–409

Milliman JD (1993) Production and accumulation of calcium-carbonate in the ocean–budget of a nonsteady state. Glob Biogeochem Cycles 7:927–957

Mills E (2005) Insurance in a climate of change. Science 309:1040–1044

Monnin E, Indermühle A, Dällenbach A, Flückiger J, Stauffer B, Stocker TF, Raynaud D, Barnola J-M (2001) Atmospheric CO_2 concentrations over the last glacial termination. Science 291:112–114

Moore JK, Doney SC, Glover DM, Fung IY (2002) Iron cycling and nutrient-limitation patterns in surface waters of the world ocean. Deep-Sea Res II 49:463–507

Nausch G, Nehring D, Nagel K (2008) Nutrient concentrations, trends and their relation to eutrophication. In: Feistel R, Nausch G, Wasmund N (eds) State and evolution of the Baltic Sea, 1952–2005. Wiley, Hoboken, pp 337–393

New M, Hulme M, Jones PD (1999) Representing twentieth century space—time climate variability part I: development of a 1961–1990 mean monthly terrestrial climatology. J Clim 12:829–856

Nicholls RJ, Wong PP, Burkett VR, Codignotto JO, Hay JE, McLean RF, Ragoonaden S, Woodroffe CD (2007) Coastal systems and low-lying areas. In: Parry ML, Canziani OF, Palutikof JP, van der Linden PJ, Hanson CE (eds) Climate change: impacts, adaptation and vulnerability: contribution of working group II to the fourth assessment report of the intergovernmental panel on climate change. Cambridge University Press, Cambridge, pp 315–356

Noiri Y, Kudo I, Kiyosawa H, Nishioka J, Tsuda A (2005) Influence of iron and temperature on growth, nutrient utilization ratios and phytoplankton species composition in the western subarctic Pacific Ocean during the SEEDS experiment. Prog Oceanogr 64:149–166

Ohlson M (1990) Some aspects of a budget for total carbonate in the Baltic Sea. In: Proceedings of the 17th conference of the Baltic oceanographers, Norrköping, pp 78–83

Omstedt A, Axell LB (2003) Modeling the variations of salinity and temperature in the large Gulfs of the Baltic Sea. Cont Shelf Res 23:265–294

Omstedt A, Elken J, Lehmann A, Piechura J (2004) Knowledge of the Baltic Sea physics gained during the BALTEX and related programmes. Prog Oceanogr 63:1–28

Omstedt A, Gustafsson E, Wesslander K (2009) Modelling the uptake and release of carbon dioxide in the Baltic Sea surface water. Cont Shelf Res 29:870–885

Orr JC, Fabry VJ, Aumont O, Bopp L, Doney SC, Feely RA, Gnanadesikan A, Gruber N, Ishida A, Joos F, Key RM, Lindsay K, Maier-Reimer E, Matear R, Monfray P, Mouchet A, Najjar RG, Plattner G-K, Rodgers KB, Sabine CL, Sarmiento JL, Schlitzer R, Slater RD, Totterdell IJ, Weirig M-F, Yamanaka Y, Yool A (2005) Anthropogenic ocean acidification over the twenty-first century and its impact on calcifying organisms. Nature 437:681–686

Park PK, Gordon LI, Hager SW, Cissell MC (1969) Carbon dioxide partial pressure in the Columbia River. Science 166:867–868

Peel MC, Finlayson BL, McMahon TA (2007) Updated world map of the Köppen–Geiger climate classification. Hydrol Earth Syst Sci 11:1633–1644

Pempkowiak J (1985) The input of biochemically labile and resistant organic matter to the Baltic Sea from the Vistula River. In: Degens ET, Kempe S, Herrera R (eds) Transport of carbon and minerals in major world rivers, SCOPE/UNEP Sonderband vol 58, part 3. Mitt. Geol.-Paläont. Inst. Univ. Hamburg, pp 345–350

Pempkowiak J, Kupryszewski G (1980) The input of organic matter to the Baltic from the Vistula River. Oceanologia 12:80–98

Pempkowiak J, Pocklington R (1983) Phenolic aldehydes as indicators of the origin of humic substances in marine environments. In: Christman RF, Gjessing ET (eds) Aquatic and terrestrial humic materials. Ann Arbor Science, Michigan, pp 371–385

Pempkowiak J, Widrowski M, Kuliński W (1984) Dissolved organic carbon and particulate carbon in the Southern Baltic in September, 1983. In: Proceedings XIV conference of Baltic oceanographers. IMGW, Gdynia, Poland, pp 699–713

Pempkowiak J, Bełdowski J, Kuliński K (2009) Zmiany globalne a zagrożenia dla, i ze strony Morza Bałtyckiego (Global changes and threat for and from the Baltic Sea). In: Gutry-Korycka M, Markowski T (eds) Zrównoważone warunki życia w zmieniającym się systemie klimatycznym Ziemi. Studia Tom CXXIV. Komitet Przestrzennego Zagospodarowania Kraju PAN, Warszawa, pp 137–149 (in Polish)

Penner JE, Andreae M, Annegarn H, Barrie L, Feichter J, Hegg D, Jayaraman A, Leaitch R, Murphy D, Nganga J, Pitari G (2001) Aerosols, their direct and indirect effects. In: Houghton JT, Ding Y, Griggs DJ, Noguer M (eds) Climate change: the scientific basis: contribution of working group I to the third assessment report of the intergovernmental panel on climate change. Cambridge University Press, Cambridge, pp 289–348

Penner JE, Zhang SY, Chuang CC (2003) Soot and smoke aerosol may not warm climate. J Geophys Res 108:D214657

Penner JE, Dong X, Chen Y (2004) Observational evidence of a change in radiative forcing due to the indirect aerosol effect. Nature 427:231–234

Petit JR, Jouzel J, Raynaud D, Barkov NI, Barnola J-M, Basile I, Bender M, Chappellaz J, Davis M, Delaygue G, Delmotte M, Kotlyakov VM, Legrand M, Lipenkov VY, Lorius C, Pepin L, Ritz C, Saltzman E, Stievenard M (1999) Climate and atmospheric history of the past 420, 000 years from the Vostok ice core, Antarctica. Nature 399:429–436

Petterson C, Allard B, Boren H (1997) River discharge of humic substances and humic-bound metals to the Gulf of Bothnia. Estuar Coast Shelf Sci 44:533–541

Pocklington R, Pempkowiak J (1984) Contribution of humic substances by the Vistula River to the Baltic Sea. In: Degens ET, Kempe S, Soliman H (eds) Transport of carbon and minerals in major world rivers, vol 55, part 2. SCOPE/UNEP Sonderband, Hamburg, pp 365–370

Prowe AEF, Thomas H, Pätsch J, Kühn W, Bozec Y, Schiettecatte L-S, Borges AV, de Baar HJW (2009) Mechanisms controlling the air–sea CO_2 flux in the North Sea. Cont Shelf Res. doi:10.1016/j.csr.2009.06.003

Quay P (2002) Ups and downs of CO_2 uptake. Science 298:2344

Rachold V, Eicken H, Gordeev VV, Grigoriev MN, Hubberten H-W, Lisitzin A P, Shevchenko V P, Schirrmeister L (2004) Modern terrigenous organic carbon input to the Arctic Ocean. In: Stein R, Macdonald RW (eds) The organic carbon cycle in the Arctic Ocean. Springer, Berlin, pp 33–55

Rahmstorf S (2002) Ocean circulation and climate during the past 120, 000 years. Nature 419:207–214

Rahmstorf S (2003) Thermohaline circulation: the current climate. Nature 421:699

Riebesell U, Schulz KG, Bellerby RGJ, Botros M, Fritsche P, Meyerhöfer M, Neill C, Nondal G, Oschlies A, Wohlers J, Zöllner E (2007) Enhanced biological carbon consumption in a high CO_2 ocean. Nature 450:545–548

Rind D (2002) The sun's role in climate variability. Science 296:673–677

Roberts GC, Ramana MV, Corrigan C, Kim D, Ramanathan V (2008) Simultaneous observations of aerosol-cloud-albedo interactions with three stacked unnamed aerial vehicles. Proc Natl Acad Sci U S A 105:7370–7375

Robock A (2000) Volcanic eruptions and climate. Rev Geophys 38:191–219

Rosenzweig C (1994) Maize suffers a sea-change. Nature 370:175–176

Ruddiman WF (2003) The anthropogenic greenhouse era began thousands of years ago. Clim Change 61:261–293

Ruddiman WF (2005) How did humans first alter global climate? Sci Am 292:34–41

Ruddiman WF, Vavrus SJ, Kutzbach JE (2005) A test of the overdue-glaciation hypothesis. Quat Sci Rev 24:1–10

Rullkötter J (2006) Organic matter: the driving force for early diagenesis. In: Schulz HD, Zabel M (eds) Marine geochemistry. Springer, Berlin, pp 125–206

Sabine CL, Feely RA, Gruber N, Key RM, Lee K, Bullister JL, Wanninkhof R, Wong CS, Wallace DWR, Tilbrook B, Millero FJ, Peng T-H, Kozyr A, Ono T, Rios AF (2004) The ocean sink for anthropogenic CO_2. Science 305:367–371

Sarmiento JL, Gruber N (2006) Ocean biogeochemical dynamics. Princeton University Press, Princeton, p 503

Sarmiento JL, Hughes TMC, Stouffer RJ, Manabe S (1998) Simulated response of the ocean carbon cycle to anthropogenic climate warming. Nature 393:245–249

Sasai Y, Ikeda M (2003) A model study for the carbon cycle in the upper layer of the North Pacific. Mar Chem 81:71–88

Sayles FL, Eck C (2009) An autonomous instrument for time series analysis of TCO_2 from oceanographic moorings. Deep-Sea Res I 56:1590–1603

Schaap M, van Loon M, ten Brink HM, Dentener FJ, Builtjes PJH (2004) Secondary inorganic aerosol simulations for Europe with special attention to nitrate. Atmos Chem Phys 4:857–874

Schmidt-Thomé P, Viehhauser M, Staudt M (2006) A decision frame for climate change impacts on sea level and river runoff: case studies of the Stockholm and Gdansk areas in the Baltic Sea region. Quat Int 145–146:135–144

Schneider B, Nagel K, Struck U (2000) Carbon fluxes across the halocline in the eastern Gotland Sea. J Mar Syst 25:261–268

Schulze E-D, Heinze C, Gash J, Volbers A, Freibauer A, Kentarchos A (2009) Integrated assessment of the European and North Atlantic carbon balance—key results, policy implications for post 2012 and research need. Office for Official Publications of the European Communities, Brussels, p 143

Schuster U, Watson AJ, Bates NR, Corbiere A, Gonzalez-Davila M, Metzl N, Pierrot D, Santana-Casiano M (2009) Trends in North Atlantic sea-surface fCO_2 from 1990 to 2006. Deep-Sea Res II 56:620–629

Siegenthaler U, Joos F (1992) Use of a simple model for studying oceanic tracer distributions and the global carbon cycle. Tellus 44:186–207

Stein R, Macdonald RW (2004) Organic carbon budget: Arctic Ocean versus Global Ocean. In: Stein R, Macdonald RW (eds) The organic carbon cycle in the Arctic Ocean. Springer, Berlin, pp 315–363

Stigebrandt A (2001) Physical oceanography of the Baltic Sea. In: Wulff FV, Rahm LA, Larsson P (eds) A system analysis of the Baltic Sea. Springer, Berlin, pp 19–74

Stigebrandt A, Gustafsson BG (2003) Response of the Baltic Sea to climate change—theory and observations. J Sea Res 49:243–256

Stocker TF (2002) North–south connections. Science 297:1814–1815

Strassmann KM, Joos F, Fischer G (2008) Simulating effects of land use changes on carbon fluxes: past contributions to atmospheric CO_2 increases and future commitments due to losses of terrestrial sink capacity. Tellus 60B:583–603

Struck U, Emeis KC, Voss M, Christiansen C, Kunzendorf H (2000) Records of southern and central Baltic Sea eutrophication in $\delta^{13}C$ and $\delta^{15}N$ of sedimentary organic matter. Mar Geol 164:157–171

Struck U, Pollehne F, Bauerfeind E, von Bodungen B (2004) Sources of nitrogen for the vertical particle flux in the Gotland Sea (Baltic proper)—results from sediment trap studies. J Mar Syst 45:91–101

Sulzman EW (2007) Stable isotopes chemistry and measurement: a primer. In: Michener R, Lajtha K (eds) Stable isotopes in ecology and environmental science. Blackwell, Singapore, 1–21

Szczepańska A, Maciejewska A, Zaborska A, Kuliński K, Pempkowiak J (2011) Distribution and provenience of organic carbon in the Baltic sediments. Geochronometria (submitted)

Takahashi T, Sutherland SC, Sweeney C, Poisson A, Metzl N, Tilbrook B, Bates N, Wanninkhof R, Feely RA, Sabine C, Olafson J, Nojiri Y (2002) Global sea–air CO_2 flux based on climatological surface ocean pCO_2, and seasonal biological and temperature effects. Deep-Sea Res II 49:1601–1622

Tegen I, Fung I (1995) Contribution to the atmospheric mineral aerosol load from land surface modification. J Geophys Res 100:18707–18726

Thomas H, Schneider B (1999) The seasonal cycle of carbon dioxide in Baltic Sea surface waters. J Mar Syst 22:53–67

Thomas H, Pempkowiak J, Wulff F, Nagel K (2003) Autotrophy, nitrogen accumulation and nitrogen limitation in the Baltic Sea: a paradox or a buffer for eutrophication? Geophys Res Lett 30:GL017937

Thomas H, Bozec Y, Elkalay K, de Baar HJW (2004) Enhanced open ocean storage of CO_2 from Shelf Sea pumping. Science 304:1005–1007

Thomas H, Bozec Y, de Baar HJW, Elkalay K, Frankignoulle M, Schiettecatte L-S, Kattner G, Borges AV (2005) The carbon budget of the North Sea. Biogeosciences 2:87–96

Thomas H, Schiettecatte L-S, Suykens K, Koné YJM, Shadwick EH, Prowe AEF, Bozec Y, de Baar HJW, Borges AV (2009) Enhanced ocean carbon storage from anaerobic alkalinity generation in coastal sediments. Biogeosciences 6:267–274

Torres R, Ampuero P (2009) Strong CO_2 outgassing from high nutrient low chlorophyll coastal waters off central Chile (30°S): the role of dissolved iron. Estuar Coast Shelf Sci 83:126–132

Trenberth KE, Caron JM (2001) Estimates of meridional atmosphere and ocean heat transports. J Clim 14:3433

Trenberth KE, Jones PD, Ambenje P, Bojariu R, Easterling D, Klein Tank A, Parker D, Rahimzadeh F, Renwick JA, Rusticucci M, Soden B, Zhai P (2007) Observations: surface and atmospheric climate change. In: Solomon S, Qin D, Manning M, Chen Z, Marquis M, Averyt KB, Tignor M, Miller HL (eds) Climate Change 2007: the physical science basis: contribution of working group I to the fourth assessment report of the intergovernmental panel on climate change. Cambridge University Press, Cambridge, pp 235–336

Tsunogai S, Watanabe S, Satao T (1999) Is there a "continental shelf pump" for the absorption of atmospheric CO_2? Tellus B 5:701–712

Twomey S (1974) Pollution and planetary albedo. Atmos Environ 8:1251–1256

Ulbrich U, Brücher T, Fink AH, Leckebusch GC, Krüger A, Pinto JG (2003) The central European floods of August 2002: part 1: rainfall periods and flood development. Weather 58:371–377

Velders GJM, Madronich S, Clerbaux C, Derwent R, Grutter M, Hauglustaine D, Incecik S, Ko M, Libre J-M, Nielsen O, Stordal F, Zhu T (2005) Chemical and radiative effects of halocarbons and their replacement compounds. In: Metz B, Kuijpers L, Solomon S, Andersen SO, Davidson O, Pons J, de Jager D, Kestin T, Manning M, Meyer LA (eds) Special report on safeguarding the ozone layer and the global climate system: issues related to hydrofluorocarbons and perfluorocarbons. Cambridge University Press, Cambridge, pp 133–180

Vichi M, Ruardij P, Baretta JW (2004) Link or sink: a modelling interpretation of the open Baltic biogeochemistry. Biogeosciences 1:79–100

Voipio A (1981) The Baltic Sea. Elsevier Scientific Publishing Company, Amsterdam, p 418

Wanninkhof R, McGillis WR (1999) A cubic relationship between air-sea CO_2 exchange and windspeed. Geophys Res Lett 26:1889–1892

Wasmund N, Siegel H (2008) Phytoplankton. In: Feistel R, Nausch G, Wasmund N, (eds) State and evolution of the Baltic Sea, 1952–2005. Wiley, Hoboken, pp 441–481

Wasmund N, Uhlig S (2003) Phytoplankton trends in the Baltic Sea. J Mar Syst 60:177–186

Wasmund N, Andrushaitis A, Łysiak-Pastuszak E, Müller-Karulis B, Nausch G, Neumann T, Ojaveer H, Olenina I, Postel L, Witek Z (2001) Trophic status of the south-eastern Baltic Sea: a comparison of coastal and open areas. Estuar Coast Shelf Sci 53:849–864

Watson AJ, Orr JC (2003) Carbon dioxide fluxes in the Global Ocean. In: Fasham MJR (ed) Ocean biogeochemistry. Springer, Berlin, pp 123–143

Wei G, McCulloch MT, Mortimer G, Deng W, Xie L (2009) Evidence for ocean acidification in the Great Barrier Reef of Australia. Geochim Cosmochim Acta 73:2332–2346

White WB (2006) Response of tropical global ocean temperature to the Sun's quasi-decadal UV radiative forcing of the stratosphere. J Geophys Res 111:C09020

Worrell E, Price L, Martin N, Hendriks C, Meida LO (2001) Carbon dioxide emissions from the global cement industry. Annu Rev Energy Environ 26:303–329

Wulff F, Rahm L, Hallin A-K, Sandberg J (2001) A nutrient budget model of the Baltic Sea. In: Wulff FV, Rahm LA, Larsson P (eds) A system analysis of the Baltic Sea. Springer, Berlin, pp 354–372

Wyrzykowski T (1985) Map of the recent vertical movements of the surface of the earth crust on the territory of Poland. Instytut Geodezjii i Kartografii, Warszawa, p 15

Zsolnay A (1973) Distribution of labile and residual particulate carbon in the Baltic Sea. Mar Biol 21:13–18

Chapter 3
Research on Carbon Cycling in the Baltic: Aims and Scope

3.1 Aims of the Studies

The main goal of the study is to present a quantitative and qualitative description of the Baltic Sea carbon cycle. In order to achieve the goal, a quantitative annual budget of the basin carbon inflows and outflows was constructed. The balance was then used to determine the direction and extent of CO_2 exchange between the seawater and the atmosphere for the entire Baltic Sea surface.

To achieve the objectives, numerous research tasks, including both empirical and mathematical research, have been implemented. While general focus is being given to the Baltic Sea carbon cycle, we will not elaborate on methodical details here. To reach the objectives, the following tasks have been undertaken.

- Identification of the most important sources and sinks of carbon in the Baltic Sea;
- Estimation of carbon exchange between the Baltic Sea and the North Sea, with resolution sufficient to recognize the seasonal variability;
- Estimation of carbon load carried with river runoff to the Baltic Sea;
- Assessment of carbon deposition to the Baltic Sea sediments;
- Determination of the carbon return flux from the sediments to the pelagic zone.

3.2 The Concept of the Study: Quantitative Assessment of the Baltic Sea Carbon Cycle

3.2.1 The "Box Model"

Quantitative approach to the carbon cycle in a marine environment has become a standard with increasing interest in mechanisms underlying the role of seas and

K. Kuliński and J. Pempkowiak, *Carbon Cycling in the Baltic Sea*,
Geoplanet: Earth and Planetary Sciences, DOI: 10.1007/978-3-642-19388-0_3,

oceans in the cycle of anthropogenic CO_2. Numerous publications present global estimations of carbon accumulated within various compartments of the marine environment as well as the magnitude of the carbon fluxes between these compartments (Sects. 2.2.1 and 2.2.2; Doney et al. 2003; Sarmiento and Gruber 2006; Emerson and Hedges 2008). These estimations allow to identify particular forcings and their impact on the carbon cycle constituents.

Carbon cycle studies performed on a local scale have led to better understanding of the marine environment and particular basins contribution to the global processes (Stein and Macdonald 2004; Borges 2005; Chen and Borges 2009). In the case of the Baltic Sea, quantitative descriptions of the carbon cycle are fragmentary and relate to selected regions and forms of carbon (Pempkowiak 1985; Algesten et al. 2006; Thomas et al. 2003). Unlike the studies on biogenic substances (e.g. Wulff et al. 2001), there is a lack of studies that describe the Baltic Sea in a more comprehensive way.

In terms of direction and magnitude of CO_2 exchange through the seawater/atmosphere interface is the most controversial/interesting among all considered carbon fluxes (Sects. 2.2.4 and 2.3.3.4). The most commonly used method to estimate its values is the ΔpCO_2 assessment, i.e., assessment of the CO_2 partial pressures difference between water and the atmosphere. However, these are usually single point measurements, and even if the data are collected frequently, such an approach does not give a complete information regarding the basin as a whole. This argument gains its significance especially for highly dynamic basins like the Baltic, characterised by both spatial and temporal changes of CO_2 concentrations (Sect. 2.3.2.4; Borges 2005; Algesten et al. 2006; Kuss et al. 2006). As a result, to calculate carbon inflows and outflows, the mass balance approach becomes the method of choice (Anderson et al. 1998; Chen et al. 2003; Borges 2005; Chen and Borges 2009). Results obtained using this method enable also to conclude on the basin trophy. Autotrophic water bodies absorb atmospheric CO_2, while heterotrophic ones are its source (Thomas et al. 2003; Borges 2005; Emerson and Hedges 2008).

These considerations were a driving force leading to the development of a complex, quantitative description of the Baltic Sea carbon cycle. The approach is based on a box model (Fig. 3.1). The basic assumption of the model is a steady state of carbon cycling—sums of all carbon inflows and outflows to and from the Baltic Sea are assumed constant (Eq. 3.1). Assuming that outflows take negative values and inflows—positive, the sum of all fluxes equals zero (Eq. 3.2). With the all but one carbon fluxes being correctly evaluated, Equation 3.3 allows to determine the missing one, e.g., the CO_2 exchange through the Baltic/atmosphere interface.

In order to construct a consistent box model, its final components are presented at the annual time scale, irrespective of the constituents dynamics. Appropriate averaging is used when necessary. Such an approach is essential due to seasonal dynamics of particular carbon fluxes (Chen et al. 2003; Doney et al. 2003).

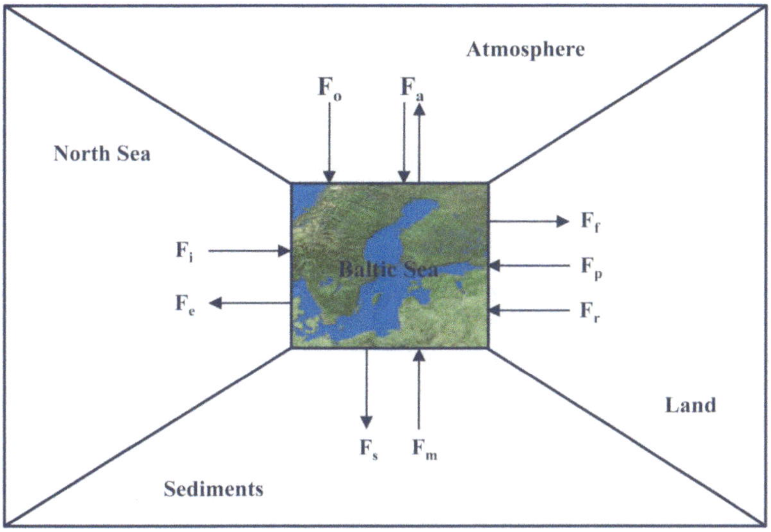

Fig. 3.1 Scheme of carbon sources and sinks considered in the Baltic Sea budget. Symbols: F_e, export to the North Sea; F_i, import from the North Sea; F_o, precipitation; F_a, net CO_2 exchange through the seawater/atmosphere interface; F_f, fisheries; F_p, point sources; F_r, rivers; F_m, return flux from the sediments; F_s, sedimentation

$$\sum \text{carbon sinks} = \sum \text{carbon sources} \tag{3.1}$$

$$F_e + F_i + F_o + F_a + F_f + F_p + F_r + F_m + F_s = 0 \tag{3.2}$$

$$F_a = F_e + F_i + F_o + F_f + F_p + F_r + F_m + F_s \tag{3.3}$$

3.2.2 Significant Carbon Fluxes in the Baltic Sea

Among the considered carbon fluxes (Fig. 3.1), the following are of particular interest: net CO_2 exchange through the seawater/atmosphere boundary, carbon exchange between the Baltic and the North Sea, carbon burial in sediments and carbon inflow with river runoff. When the CO_2 flux through the water/atmosphere boundary is determined using the mass balance method, the considered carbon fluxes become decisive for the direction and intensity of the exchange. These conclusions are based on the literature review of the subject, described in Sect. 2.3.3, which conveys information about the potential of particular constituents of carbon budget in the Baltic Sea. In order to assess the correct mass balance and to estimate the direction and extent of CO_2 exchange through the water/ atmosphere boundary in the possibly most precise way, a search for the existing estimations of the above-mentioned carbon fluxes was carried out. The problems encountered in the literature studies are presented below:

- Previous estimations of the carbon exchange between the Baltic Sea and the North Sea (Sect. 2.3.3.1; Thomas et al. 2003) were based on carbon concentrations and volume of flowing water annual averages, while these values are subjected to substantial seasonal variability (Thomas and Schneider 1999; Suratman et al. 2007; Lass and Matthäus 2008; Prowe et al. 2009). Moreover, the North Sea and the Baltic Sea water masses are subjected to complex mixing processes occurring within the Danish Straits as well as at the boundary zones of the Baltic and the North Sea (Lass and Matthäus 2008).
- The current estimates regarding carbon inflow with rivers are based on a limited number of available data. This is especially explicit for DIC concentrations obtained by extrapolating the relationships of DIC concentration and salinity measured for the Baltic Sea open waters. Moreover, available results do not consider seasonal changes of carbon load transported by rivers (Sect. 2.3.3.2; Thomas et al. 2003; Algesten et al. 2006).
- Estimates of organic carbon accumulation rates in sediments are based on the sediment deposition rates derived from ^{210}Pb activity concentrations profiles and organic carbon concentrations in sediments (Sect. 2.3.3.3). Accelerated carbon accumulation rates during the last decades are attributed to an increased inflow of organic matter to sediments due to the Baltic Sea eutrophication processes (Emeis et al. 2000; Struck et al. 2000; Algesten et al. 2006). Unfortunately, these estimates do not include processes of the early diagenesis, that lead to a partial decomposition and mineralisation of the organic matter deposited to sediments and return flux of dissolved carbon to the water column (Almroth et al. 2009).

3.3 The Scope of Studies

Based on the considerations described above, the following research tasks have been determined in order to develop new methods enabling quantitative estimation and improvement of existing estimates of key processes influencing the Baltic Sea carbon budget:

- Estimation of carbon exchange between the Baltic Sea and the North Sea based on temporal resolution not worst than one week and accompanied by processes of both water masses mixing and seasonality of DIC and DOC concentrations, for the Baltic and the North Sea separately (new method—improved reliability and precision);
- Expanding data set concerning carbon discharged to the Baltic Sea with river runoff (existing method—improved accuracy);
- Quantification of carbon burial in the Baltic Sea sediments, based on carbon accumulation rates and the carbon return flux from sediments to the water

column resulting from the early diagenesis processes dynamics (new method—improved reliability and accuracy).

References

Algesten G, Brydsten L, Jonsson P, Kortelainen P, Löfgren S, Rahm L, Räike A, Sobek S, Tranvik L, Wikner J, Jansson M (2006) Organic carbon budget for the Gulf of Bothnia. J Mar Syst 63:155–161

Almroth E, Tengberg A, Adersson JH, Pakhomova S, Hall POJ (2009) Effects of resuspension on the benthic fluxes of oxygen, nutrients, dissolved inorganic carbon, iron and manganese in the Gulf of Finland, Baltic Sea. Cont Shelf Res 29:807–818

Anderson LG, Olsson K, Chierici M (1998) A carbon budget for the Arctic Ocean. Global Biogeochem Cycles 12:455–465

Borges AV (2005) Do we have enough pieces of the jigsaw to integrate CO_2 fluxes in the coastal ocean? Estuaries 28:3–27

Chen C-TA, Borges AV (2009) Reconciling opposing views on carbon cycling in the coastal ocean: continental shelves as sinks and near-shore ecosystems as sources of atmospheric CO_2. Deep-Sea Res II 56:578–590

Chen C-TA, Liu K-K, Macdonald R (2003) Continental margin exchanges. In: Fasham MJR (ed) Ocean biogeochemistry. Springer, Berlin, pp 53–97

Doney SC, Lindsay K, Moore JK (2003) Global ocean carbon cycle modeling. In: Fasham MJR (ed) Ocean biogeochemistry. Springer, Berlin, pp 217–238

Emeis K-C, Struck U, Leipe T, Pollehne F, Kunzendorf H, Christiansen C (2000) Changes in the C, N, P burial rates in some Baltic Sea sediments over the last 150 years–relevance to P regeneration rates and the phosphorus cycle. Mar Geol 167:43–59

Emerson SR, Hedges JI (2008) Chemical oceanography and the marine carbon cycle. Cambridge University Press, Cambridge, p 453

Kuss J, Roeder W, Wlost KP, DeGrandpre MD (2006) Time-series of surface water CO_2 and oxygen measurements on a platform in the central Arkona Sea (Baltic Sea): seasonality of uptake and release. Mar Chem 101:220–232

Lass H-U, Matthäus W (2008) General oceanography of the Baltic Sea. In: Feistel R, Nausch G, Wasmund N (eds) State and evolution of the Baltic Sea, 1952–2005. Wiley, Hoboken, pp 5–43

Pempkowiak J (1985) The input of biochemically labile and resistant organic matter to the Baltic Sea from the Vistula River. In: Degens ET, Kempe S, Herrera R (eds) Transport of carbon and minerals in major world rivers, Pt. 3., Mitt. Geol.- Paläont. Inst. Univ. Hamburg, SCOPE/UNEP Sonderband 58, 345–350

Prowe AEF, Thomas H, Pätsch J, Kühn W, Bozec Y, Schiettecatte L-S, Borges AV, de Baar HJW (2009) Mechanisms controlling the air-sea CO_2 flux in the North Sea. Cont Shelf Res 29:1801–1808

Sarmiento JL, Gruber N (2006) Ocean biogeochemical dynamics. Princeton University Press, Princeton, p 503

Stein R, Macdonald RW (2004) Organic carbon budget: Arctic Ocean vs. Global Ocean. In: Stein R, Macdonald RW (eds) The organic carbon cycle in the Arctic Ocean. Springer, Berlin, pp 315–363

Struck U, Emeis KC, Voss M, Christiansen C, Kunzendorf H (2000) Records of southern and central Baltic Sea eutrophication in $\delta^{13}C$ and $\delta^{15}N$ of sedimentary organic matter. Mar Geol 164:157–171

Suratman S, Jickells T, Weston K, Fernand L (2007) Seasonal changes of dissolved and particulate organic C and N in the North Sea. Geophys Res Abstr 9:03651

Thomas H, Schneider B (1999) The seasonal cycle of carbon dioxide in Baltic Sea surface waters. J Mar Syst 22:53–67

Thomas H, Pempkowiak J, Wulff F, Nagel K (2003) Autotrophy, nitrogen accumulation and nitrogen limitation in the Baltic Sea: a paradox or a buffer for eutrophication? Geophys Res lett 30, GL017937

Wulff F, Rahm L, Hallin A-K, Sandberg J (2001) A nutrient budget model of the Baltic Sea. In: Wulff FV, Rahm LA, Larsson P (eds) A system analysis of the Baltic Sea. Springer, Berlin, pp 354–372

Chapter 4
Research on Carbon Cycling in the Baltic: Quantification of the Carbon Fluxes

4.1 Carbon Exchange Between the Baltic Sea and the North Sea

Considerations presented in this chapter are based on the comprehensive study performed by Kuliński et al. (2011). The details of the methodology used for the quantification of carbon exchange between the Baltic Sea and the North Sea are presented in Sect. A.1 (this volume, Appendixes).

The key element that determines carbon exchange between the Baltic and the North Sea is the direction and volume of water flows between two basins. "End members" method allows to distinguish these two water masses within the total water volume that enters the Danish Straits and to verify the respective carbon contributions to export/import to/from the Baltic. Simultaneously, the method allows disentangling of the mixing process complexity occurring within the Danish Straits.

Annual variability of water exchange for the period from June 1, 2002, to May 31, 2006, is illustrated by series of diurnal flows of the Baltic (Fig. 4.1) and the North Sea water masses (Fig. 4.2). In both cases, negative values indicate the water flowing from the Baltic to the North Sea, while positive values indicate an opposite direction. During the entire considered period, the direction and the volume of the flowing water are subjected to significant variability. The Baltic daily water flows range from 17.1 km^3 day^{-1} (December 7, 2002) to 17.5 km^3 day^{-1} (February 13, 2005), compared to -13.3 km^3 day^{-1} (January 15, 2005) to 20.4 km^3 day^{-1} (November 15, 2006) for the North Sea.

In order to estimate the annual magnitude of carbon exchange variability between the Baltic and the North Sea, four yearly intervals have been distinguished, based on the hydrological data, for the period from June 1 to May 31 (Table 4.1). Altogether, during four years (June 1, 2002 to May 31, 2006) the North Sea received 5658.6 km^3 of water from the Baltic. During the same period

K. Kuliński and J. Pempkowiak, *Carbon Cycling in the Baltic Sea*,
Geoplanet: Earth and Planetary Sciences, DOI: 10.1007/978-3-642-19388-0_4,

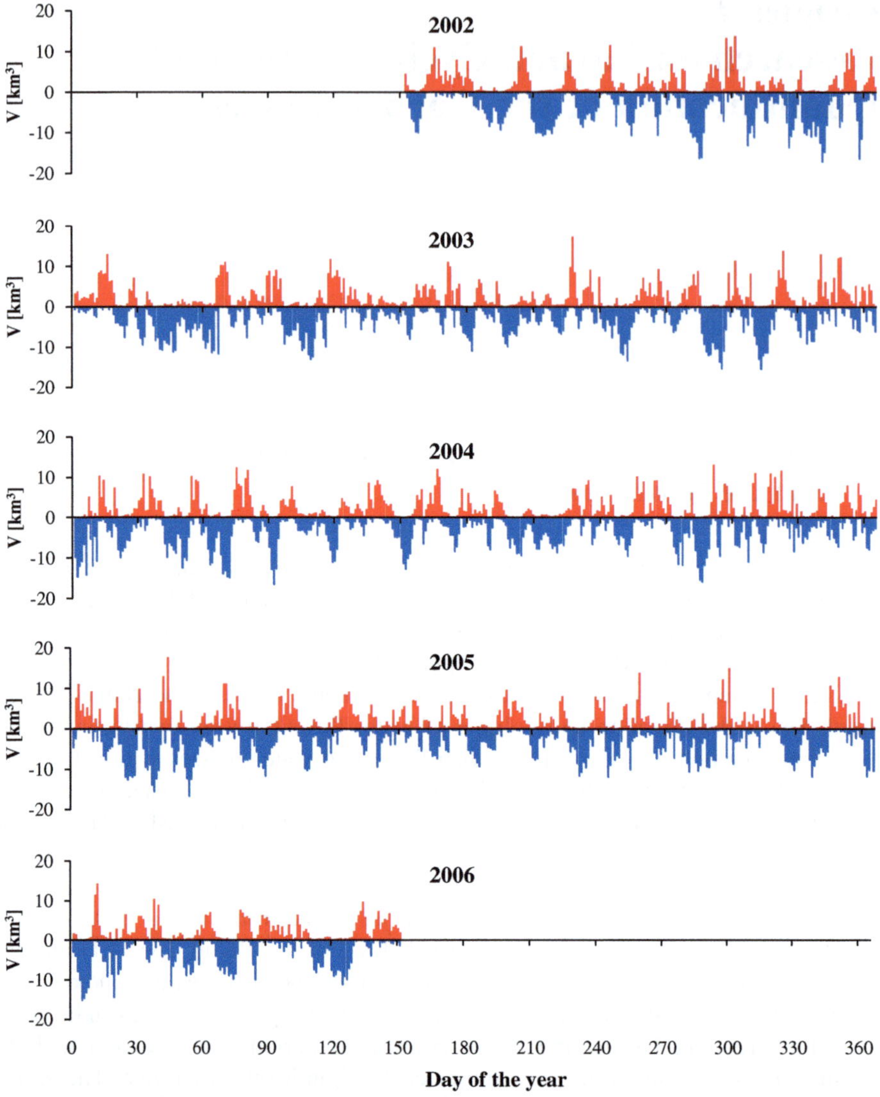

Fig. 4.1 Seasonal variations of the daily Baltic water flows during the period from June 1, 2002, to May 31, 2006, calculated based on the data generated by CMOD hydrological model and the "end members" method. Negative values (*blue*) indicate the Baltic water flowing into the North Sea, positive values (*red*) indicate the Baltic water returning to its source (modified after Kuliński et al. 2011)

of time, the volume of 3594.0 km^3 of the Baltic water returning to its source was recorded, which constitutes 63% of the export. Annual volumes of the Baltic water exported to the North Sea ranged from -1333.9 km^3 $year^{-1}$ for the period of

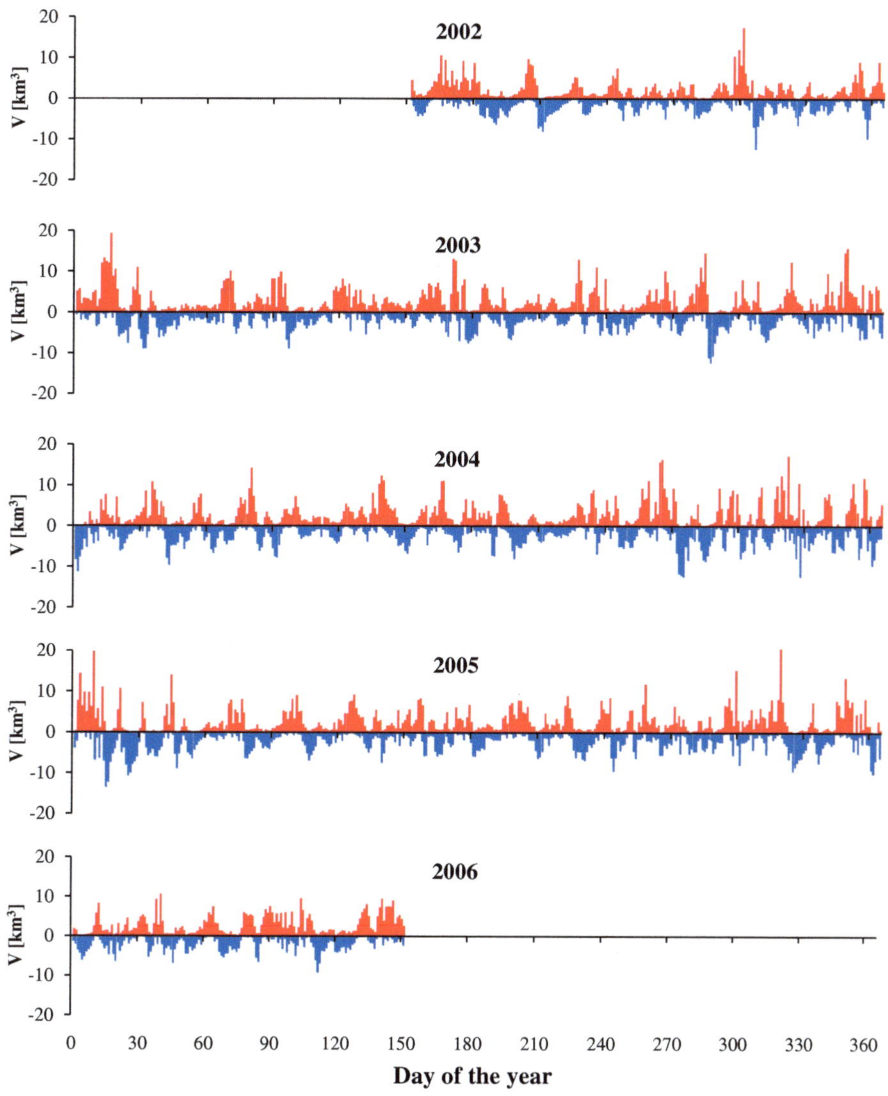

Fig. 4.2 Seasonal variations of the daily North Sea water flows during the period from June 1, 2002, to May 31, 2006, calculated based on the data generated by CMOD hydrological model and the "end members" method. Positive values (*red*) indicate the North Sea water inflows to the Baltic, negative values (*blue*) indicate the North Sea water returning to its source (modified after Kuliński et al. 2011)

2005/2006 to -1496.9 km^3 year^{-1} during 2002/2003, while volumes of the Baltic water returning to its source varied from 854.4 km^3 year^{-1} during 2005/2006 to 929.0 km^3 year^{-1} during 2004/2005. The Baltic Sea water net export to the North

Table 4.1 Water exchange between the Baltic Sea and the North Sea (Kuliński et al. 2011)

Period	Baltic water			North Sea Water		
	Exported from Baltic (km^3)	Imported to Baltic (km^3)	Net export from Baltic (km^3)	Imported to Baltic (km^3)	Exported from Baltic (km^3)	Net import to Baltic (km^3)
2002/2003	−1496.9	886.6	−610.3	965.6	−749.8	206.8
2003/2004	−1422.0	924.0	−498.0	1035.0	−876.9	158.1
2004/2005	−1405.8	929.0	−476.8	1059.0	−981.4	77.6
2005/2006	−1333.9	854.4	−479.5	1008.7	−857.1	151.6
Average	−1414.6	898.5	−516.1	1014.8	−866.3	148.5
SD[a]	66.9	35.0	63.5	40.0	95.0	53.3

[a] Standard deviation

Sea falls within a range of −476.8 km^3 $year^{-1}$ for 2004/2005 and −610.3 km^3 $year^{-1}$ for 2002/2003, which corresponds to 34% and 41% of the total volume of water transported during these years. The average value of −516.1 ± 63.5 km^3 $year^{-1}$ characterizes the Baltic water flow to the North Sea in the period of investigations.

During the entire period, from June 1, 2002, to May 31, 2006, the volume of 4068.3 km^3 of the North Sea water was transported to the Baltic (Table 4.1); in the same time 85% of this water volume (−3465.2 km^3) returned to its source. Annual volumes of the North Sea water entering the Baltic range from 965.6 km^3year^{-1} for 2002/2003 to 1059.0 km^3year^{-1} for 2004/2005, compared to averages of returning fluxes: −749.8 km^3year^{-1} for 2002/2003 and −981.4 km^3year^{-1} for 2004/2005. The net import of the North Sea water to the Baltic oscillates between 77.6 km^3year^{-1} for 2004/2005 and 206.8 km^3year^{-1} for 2002/2003. These values correspond, respectively, to 7% and 21% of the total North Sea water flowing to the Baltic for the above−mentioned periods. On the average 148.5 ± 53.3 km^3year^{-1} of the North Sea water enters the Baltic Sea.

4.1.1 Dissolved Inorganic Carbon

Seasonal fluctuations of DIC concentration for MB (the Baltic end member, see Sect. A1 for explanations) and MP (the North Sea end member, see Sect. A1 for explanations) stations are presented in Fig. 4.3. Literature data (Prowe et al. 2009; Thomas and Schneider 1999) were further confirmed by our own measurements performed during various seasons in recent years.

Both stations are characterized by explicit distinct seasonal variability of DIC concentrations, with distinct summer minimum and winter maximum. This is related to the atmospheric CO_2 solubility changes in the sea water dependant on the water temperature and the phytoplankton activity, leading to the dissolved CO_2 uptake during the photosynthesis (Sect. 2.2.3.2). DIC concentrations for MB

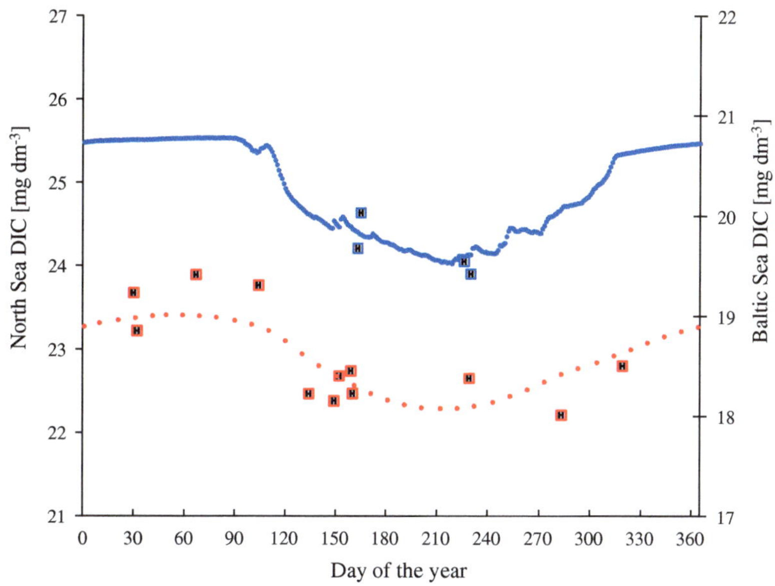

Fig. 4.3 DIC concentrations seasonal variability, curves: MP station—*blue* (Prowe et al. 2009), and MB station—*red* (Thomas and Schneider 1999). Squares—DIC concentrations based on our own data obtained during the r/v Oceania 2006–2009 research cruises (modified after Kuliński et al. 2011)

station vary from 18.1 ± 0.3 mg dm^{-3} to 19.0 ± 0.3 mg dm^{-3} estimated for 212 and 49 day of the year, respectively (July 31 and February 18). In the case of MP station, DIC concentrations oscillate at higher levels, with an amplitude of changes ranging between 24.0 ± 0.4 mg dm^{-3} for summer and 25.2 ± 0.4 mg dm^{-3} for winter (Prowe et al. 2009).

The literature data concerning DIC seasonal variations at the MB and MP stations, and daily sea water flows calculations based on the Baltic and North Sea components, enabled estimation of the annual (June 1 to May 31) DIC flows for the considered 4 years long period. The results are summarized in Table 4.2. Average Baltic DIC export to the North Sea amounts to −9.70 ± 1.13 Tg C year^{-1}, while the subsequent annual averages range from −8.98 Tg C year^{-1} for 2005/2006 to −11.39 Tg C year^{-1} for 2002/2003. The annual North Sea DIC loads imported to the Baltic are smaller and reveal more variability. Values oscillate between 1.65 Tg C year^{-1} during the period of 2004/2005 and 5.41 Tg C year^{-1} during- 2002/2003, which gives the period average equal to 3.70 ± 1.55 Tg C year^{-1} of DIC transported from the North Sea to the Baltic.

In order to illustrate seasonal variability of the magnitude and direction of DIC fluxes, including their monthly averages with standard deviation, the Baltic and North Sea DIC loads were calculated for the entire 4 year period of studies (Fig. 4.4). Only for May and June, does the Baltic DIC flux reveal positive values,

Table 4.2 Summary of the annual (June 1 to May 31) Baltic and North Sea DIC fluxes

Period	Baltic DIC (Tg C year^{-1})	North Sea DIC (Tg C year^{-1})	Total DIC (Tg C year^{-1})
2002/2003	−11.39	5.41	−5.98
2003/2004	−9.23	3.97	−5.26
2004/2005	−9.18	1.65	−7.53
2005/2006	−8.98	3.76	−5.22
average ± SD	−9.70 ± 1.13	3.70 ± 1.55	−6.00 ± 1.08

Negative values indicate DIC export from the Baltic to the North Sea, positive values indicate DIC import from the North Sea to the Baltic (Kuliński et al. 2011)

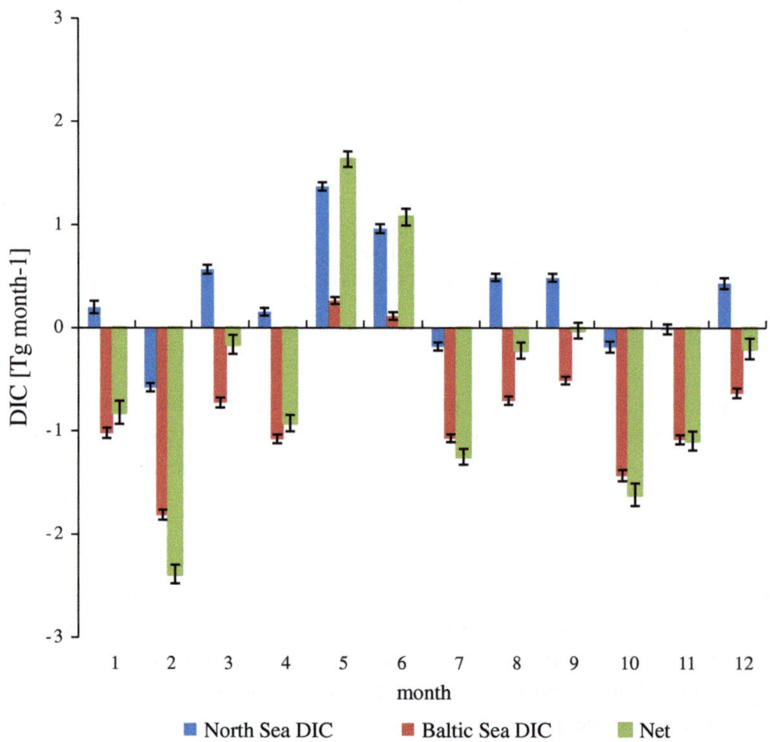

Fig. 4.4 Monthly averages of the Baltic and North Sea DIC loads and net load values for the period from June 1, 2002, to May 31, 2006. Negative values indicate DIC export from the Baltic, positive values indicate DIC import from the North Sea to the Baltic (modified after Kuliński et al. 2011)

0.26 ± 0.03 Tg C month^{-1} and 0.11 ± 0.04 Tg C month^{-1}, respectively. This indicates an intensive inorganic carbon returning flux to the Baltic from the North Sea, while for the remaining months, a DIC export to the North Sea is observed.

Monthly averages of the export range from -0.51 ± 0.04 Tg C month^{-1} in September to -1.81 ± 0.05 Tg C month^{-1} in February. Despite higher DIC concentrations, the North Sea DIC flux absolute values are somewhat smaller, which has its source in smaller volumes of the North Sea water flowing into the Baltic. The most intensive DIC fluxes from the North Sea, illustrated by monthly averages, are recorded for May and June, amounting to 1.37 ± 0.04 Tg C month^{-1} and 0.96 ± 0.04 Tg C month^{-1} respectively. During the subsequent 6 months (January, March, April, August, September, December), the DIC import from the North Sea to the Baltic is also observed, amounting to the monthly averages ranging from 0.15 ± 0.04 Tg C month^{-1} in April to 0.56 ± 0.04 Tg C month^{-1} in March. Results obtained for the remaining months indicate the North Sea DIC loads returning to its source, ranging from -0.01 ± 0.05 to -0.58 ± 0.04 Tg C month^{-1}. Monthly average totals of DIC flux presented in Fig. 4.3 as net values reveal positive values only for the month of May and June (similarly to the Baltic DIC flux). This fact clearly points to the Baltic DIC export as the determining factor for the inorganic carbon exchange between the Baltic and the North Sea. Net monthly averages of DIC exchange between basins range from -2.39 ± 0.09 Tg C month^{-1} in February to 1.63 ± 0.07 Tg C month^{-1} in May.

4.1.2 Dissolved Organic Carbon

The DOC concentrations measured in the near-shore zone, results of analysis of samples collected at both the MP and MB stations, as well as the known concentrations at the MP station (Suratman et al. 2009) allowed to estimate the DOC concentrations variability for the North Sea (MP station) and for the Baltic Sea (MB station) end members (Fig. 4.5). DOC concentrations at all stations exhibit a seasonal trend that is opposite to the DIC one (Sect. 4.1.1), characterized here by summer maximum and winter minimum. This situation is related to factors determining DOC concentrations in sea water.

DOC concentrations in the nearshore zone are characterized by significant seasonal fluctuations (Fig. 4.5). The amplitude of these changes ranges between 3.1 ± 0.2 mg dm^{-3} in the winter time and 5.0 ± 0.7 mg dm^{-3} in the summer—July and August. The estimates are charged with relative standard deviation extending from 2.3% for June 29 to 11.5% for November 8. DOC concentrations at the MB station during the winter season are comparable to those observed for the nearshore station, indicating most likely similar concentrations of refractory DOC for both regions. Seasonal changes of DOC concentrations at the MB station range from 3.1 ± 0.2 to 4.3 ± 0.2 mg dm^{-3}.

In the case of the North Sea, little dynamics of DOC concentrations during a year span was recorded (Fig. 4.5), with values ranging from 1.1 ± 0.1 mg dm^{-3} to 1.6 ± 0.1 mg dm^{-3}. This is related to lower phytoplankton activity than that in the Baltic, as well as significant influence from the North Atlantic, which shapes the North Sea water chemistry (Thomas et al. 2005).

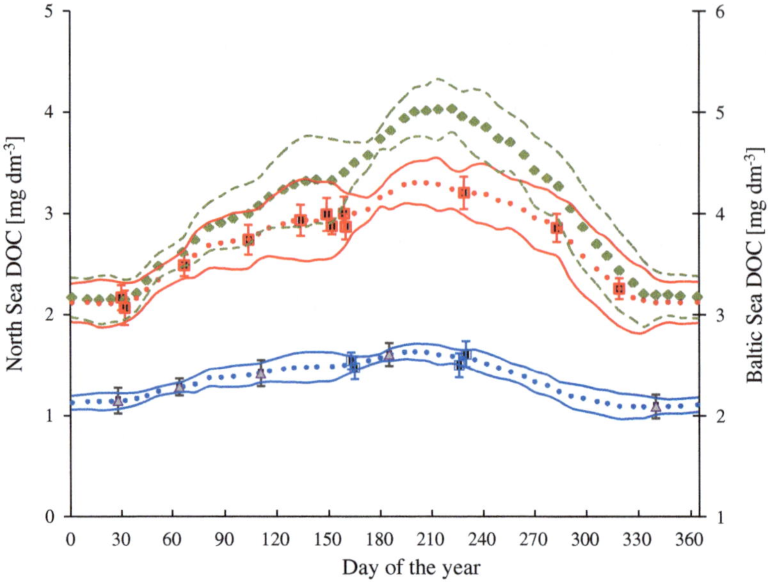

Fig. 4.5 The DOC concentrations variability: measured concentrations (*green rhombuses*), concentrations extrapolated to the MB station (*red dots*) and to the MP station (*blue dots*). Squares represent the DOC concentrations measured at MB and MP stations, triangles—DOC data from the literature (Suratman et al. 2009). *Dashed lines* define the range of the measured DOC uncertainty, whilst *solid lines*—the range of the extrapolated DOC concentrations uncertainties as given by RSD (modified after Kuliński et al. 2011)

The seasonal variability of DOC concentrations measured for MB and MP stations and calculations of daily Baltic and North Sea seawater flows, allowed to determine the annual DOC fluxes for the considered period of time. The obtained results are summarized in Table 4.3. Net values of the Baltic DOC export to the North Sea fall within -1.72 Tg C year^{-1} for 2005/2006 and -2.21 Tg C year^{-1} for 2002/2003, with the average of -1.88 ± 0.22 Tg C year^{-1}. Annual loads of the North Sea DOC entering the Baltic are definitely lower, oscillating between

Table 4.3 Summary of the annual (June 1 to May 31) Baltic and North Sea DOC fluxes

Period	Baltic DOC (Tg C year^{-1})	North Sea DOC (Tg C year^{-1})	DOC Total (Tg C year^{-1})
2002/2003	−2.21	0.28	−1.93
2003/2004	−1.80	0.23	−1.57
2004/2005	−1.79	0.11	−1.68
2005/2006	−1.72	0.22	−1.50
average ± SD	−1.88 ± 0.22	0.21 ± 0.07	−1.67 ± 0.19

Negative values indicate DOC export from the Baltic to the North Sea, positive values indicate DOC import from the North Sea to the Baltic (Kuliński et al. 2011)

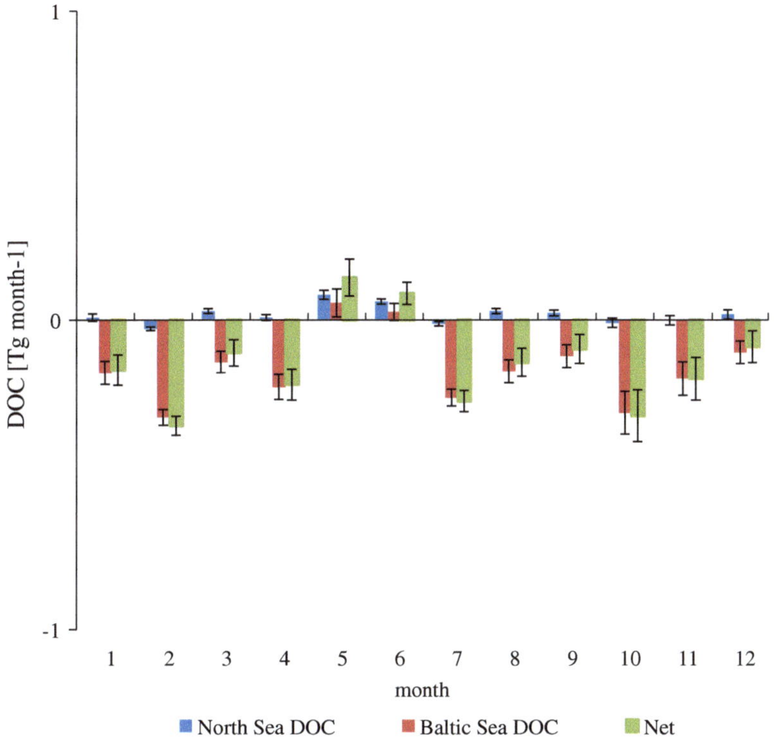

Fig. 4.6 Monthly averages of the Baltic and North Sea DOC fluxes and net flux values for the period from June 1, 2002, to May 31, 2006. Negative values indicate DOC export from the Baltic to the North Sea, positive values indicate DOC import from the North Sea to the Baltic (modified after Kuliński et al. 2011)

0.11 Tg C year^{-1} during 2004/2005 and 0.28 Tg C year^{-1} during 2002/2003. The total of the Baltic and the North Sea DOC loads, expressing quantitatively the DOC net exchange between the two basins, vary from -1.50 Tg C year^{-1} for 2005/2006 to -1.93 Tg C year^{-1} for 2003/2004, with the average of -1.67 ± 0.19 Tg C year^{-1}.

The monthly averages for the Baltic and the North Sea DOC fluxes, as illustrated in Fig. 4.6, indicate a dominant role of the former in shaping the organic carbon exchange between the two basins. The North Sea DOC load is less important and its values fall within standard deviation of the estimated net results. Similarly to DIC fluxes, it is only during May and June that positive values for monthly averages of the Baltic DOC fluxes are observed, and amount to 0.05 ± 0.04 and 0.03 ± 0.03 Tg C month^{-1}, respectively. Monthly averages of DOC for the remaining months range from -0.10 ± 0.03 Tg C month^{-1} in December to 0.30 ± 0.07 Tg C month^{-1} in October.

4.2 Carbon Inflows with River Runoff

Considerations presented in this chapter are based on the comprehensive study performed by Kuliński and Pempkowiak (2011a). The data (monthly means of water flows and carbon concentrations) used for the computation of carbon fluxes were taken from the database created and provided by Baltic-C—a FP6 BONUS funded project. This database contains the results obtained by national monitoring programmes carried out by the Baltic Sea countries.

Summary of the total inorganic carbon (TIC) and the total organic carbon (TOC) carried by (63) major rivers flowing into the Baltic Sea is presented in Table 4.4. Annually, rivers discharge, to the Baltic, 8.78 Tg of carbon, out of which more than 62% (5.49 Tg) is represented by the inorganic fraction. Among considered rivers, fifty seven (57 or 90%) enter the Baltic from the Swedish and Finnish territory. These rivers carry only 10% (0.53 Tg) of inorganic carbon and less than 40% (1.31 Tg) of organic carbon, of the total carbon load introduced to

Table 4.4 Summary of total inorganic carbon load (TIC) and total organic carbon load (TOC) carried by major rivers flowing into the Baltic Sea (Kuliński and Pempkowiak 2011a)

L. p.	River	Country	TIC (Gg year^{-1})	TOC (Gg year^{-1})	Water volume (km^3 year^{-1})
1	Ahtava	Finland	1.2	6.9	0.5
2	Ångermanälven	Sweden	37.5	84.6	16.3
3	Ätran	Sweden	7.0	19.3	1.8
4	Aura	Finland	1.7	4.1	0.2
5	Botorpström	Sweden	1.1	2.6	0.2
6	Daugava	Latvia	1068.9	210.7	20.8
7	Delångersån	Sweden	1.0	2.6	0.4
8	Emån	Sweden	4.3	16.8	1.1
9	Eura	Finland	0.9	2.7	0.2
10	Gide	Sweden	1.0	9.2	0.9
11	Göta	Sweden	63.8	80.5	18.1
12	Helgeån	Sweden	7.3	27.0	1.5
13	Iijoki	Finland	12.9	66.5	5.7
14	Indalsälven	Sweden	61.1	57.0	13.9
15	Kalajoki	Finland	2.6	26.9	1.1
16	Kalix	Sweden	20.5	59.0	10.1
17	Kemijoki	Finland	1.3	2.1	0.3
18	Kiiminki	Finland	2.9	25.8	1.5
19	Kiskonjoki	Finland	0.7	2.7	0.2
20	Kokemaenjoki	Finland	28.0	75.0	7.0
21	Koskenkyla	Finland	1.0	1.9	0.2
22	Kuivajoki	Finland	1.3	7.5	0.5
23	Kymi Ahven	Finland	21.9	46.2	5.8

(continued)

Table 4.4 (continued)

L. p.	River	Country	TIC (Gg year^{-1})	TOC (Gg year^{-1})	Water volume (km^3 year^{-1})
24	Kymi Kokon	Finland	17.9	38.5	5.1
25	Lagan	Sweden	4.8	37.9	2.8
26	Lapuanjoki	Finland	1.7	19.1	0.9
27	Lestijoki	Finland	0.5	6.2	0.3
28	Ljungan	Sweden	15.2	19.2	3.2
29	Ljungbyån	Sweden	0.3	3.4	0.2
30	Ljusnan	Sweden	0.9	0.5	0.2
31	Lule	Sweden	31.6	44.4	16.1
32	Lyckebyån	Sweden	0.4	3.8	0.2
33	Merikarvia	Finland	0.9	11.7	0.6
34	Mörrumsån	Sweden	1.7	12.8	0.9
35	Motala	Sweden	33.0	24.3	3.2
36	Mustijoki	Finland	1.3	3.6	0.2
37	Narpionjoki	Finland	0.5	7.2	0.3
38	Narva	Estonia	384.5	190.9	12.7
39	Neman	Lithuania	609.1	123.4	11.9
40	Neva	Russia	1295.2	1209.5	77.6
41	Nissan	Sweden	3.0	24.5	1.6
42	Nyköpingsån	Sweden	5.5	6.4	0.6
43	Oder	Poland	431.8	75.5	13.1
44	Öre	Sweden	1.2	13.7	1.1
45	Oulujoki	Finland	17.4	87.3	8.9
46	Paimionjoki	Finland	2.4	2.3	0.2
47	Perhonjoki	Finland	0.6	10.1	0.6
48	Pite	Szwecja	8.9	22.4	5.5
49	Porvoonjoki	Finland	3.6	5.5	0.4
50	Pyhajoki	Finland	2.3	20.0	1.0
51	Råne	Sweden	2.1	11.7	1.4
52	Rickleån	Sweden	0.7	5.8	0.5
53	Rönneån	Sweden	5.4	4.1	0.4
54	Siikajoki	Finland	3.1	29.3	1.4
55	Simojoki	Finland	3.6	20.3	1.5
56	Sirppujoki	Finland	0.1	1.1	0.1
57	Skellefte	Sweden	6.0	8.2	3.4
58	Torne	Sweden	28.9	102.4	14.2
59	Ume	Sweden	35.9	63.8	14.9
60	Uskelanojoki	Finland	1.4	2.6	0.2
61	Vantaa	Finland	4.7	8.9	0.6
62	Virojoki	Finland	0.4	2.5	0.1
63	Vistula	Poland	1168.1	175.6	28.5
	Total		5486.6	3297.9	344.9

the basin. The decisive role is played by the continental rivers: Neva, Vistula, Daugava, Neman, Narva and Odra. The continental rivers are characterized by the largest catchment areas and greatest volume of water transported to the Baltic; Neva introduces annually 77.6 km^3 $year^{-1}$ of water and 1.29 Tg TIC and 1.21 Tg TOC to the Gulf of Finland. It represents 24% and 36% of the total inorganic and organic carbon load, respectively, entering the Baltic Sea with the considered rivers.

The major rivers entering the Baltic Sea deliver 345 km^3 of water every year, which represents 80% of the total river runoff into the basin, estimated at 428 km^3 $year^{-1}$ (Lass and Matthäus 2008). Assuming the representativeness of the carbon load carried by the considered rivers, it was estimated that every year 6.81 Tg TIC $year^{-1}$ and 4.09 Tg TOC $year^{-1}$ enter the Baltic Sea with the runoff, which amounts to the annual flux, of land-originating total carbon, equal to 10.90 Tg.

4.3 Carbon Deposition to the Baltic Sea Sediments

Considerations presented in this chapter are based on the studies performed by Kuliński (2010) and Kuliński and Pempkowiak (2011b). The details of the methodology used for the quantification of carbon deposition, mineralization and burial in the Baltic Sea sediments are presented in Sect. A.2 (this volume, Appendix).

The organic carbon deposition rates to the Baltic Sea sediments are summarized in Table 4.5. The negative values of carbon flux indicate its export from the water column to sediments. The total depositional area of the sea amounts to more than 101×10^3 km^2 (Algesten et al. 2006; Błaszczyszyn 1982; Emeis et al. 2000). Each

Table 4.5 Summary of carbon loads reaching the sediments within different depositional regions, surface areas of the regions and organic carbon accumulation rates

Depositional region	Surface area (km^2)	TOC accumulation rate (g C m^{-2} $year^{-1}$)	Annual carbon deposition to sediments (Tg $year^{-1}$)
Arkona Basin	3 200[a]	20[b]	−0.064
Bornholm Deep	13 170[a]	58[c]	−0.764
Gdansk Deep	12 150[a]	60[b]	−0.729
Gotland Deep	28 240[a,b]	40[b]	−1.130
Gulf of Riga	890[a]	40[e]	−0.036
Gulf of Finland	940[a]	40[e]	−0.038
Gulf of Bothnia	42 440[d]	24[d]	−1.018
Total	101 030	–	−3.779

Negative values indicate carbon export from the water column to sediments (Kuliński 2010)
[a] (Błaszczyszyn 1982)
[b] (Emeis et al. 2000)
[c] (Christoffersen et al. 2007)
[d] (Algesten et al. 2006)
[e] Values adapted from the Gotland Deep

year, −3.78 Tg C is deposited to the sediments within the depositional areas. Due to its largest area among the regions considered, the Gulf of Bothnia is characterized by the greatest deposition loads, equal to −1.13 Tg year^{-1}. The smallest amount of carbon is deposited within the Gulf of Riga and the Gulf of Finland (−0.04 Tg year^{-1}). Accumulation rates for the organic carbon oscillate from 20 to 24 g m^{-2} year^{-1} for the Arkona Basin and the Gulf of Bothnia, and from 58 to 60 g m^{-2} year^{-1} for the Bornholm Deep and the Gdansk Deep, respectively.

4.3.1 Carbon Return Flux from the Bottom Sediments

Figures 4.7, 4.8, 4.9, 4.10, 4.11 present DIC and DOC concentrations in pore water and the above-sediment water layer for 5 among 7 considered depositional areas in the Baltic Sea. To cover all of the depositional areas of the Baltic, results characterizing the Gotland Deep were extended for the Gulf of Riga and the Gulf of Finland. In all cases, both DIC and DOC concentrations demonstrated decreasing trends in the direction of the water/sediment interface. The dependences between concentrations and depths were approximated by linear regression plots. The strength of correlation is described by the determination coefficient R^2. The coefficient values for DIC profile, ranging from 0.88 for in the Arkona Basin (Fig. 4.7) to 0.98 for the Bornholm Deep (Fig. 4.8) and Gotland Deep (Fig. 4.10), indicate a very high level of linear correlation for DIC concentration changes with depth. Subsequent profiles are characterized by varying slope, described by directional coefficients a in equation $y = ax + b$, in other words—the gradient of

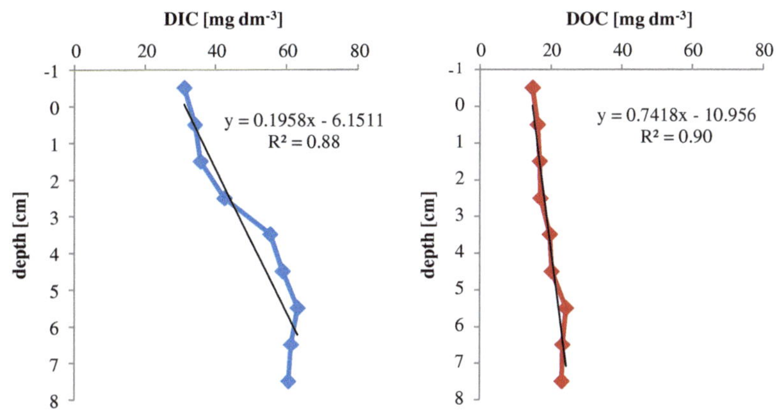

Fig. 4.7 The DIC and DOC concentration profiles in pore water from the upper layers of sediments of the Arkona Basin. The determination coefficients (R^2) indicate the linear correlaton strength. Accuracy of DIC and DOC concentrations analysis is described by relative standard deviation equal to 1.8% and 2.4%, respectively (n = 3) (modified after Kuliński and Pempkowiak 2011b)

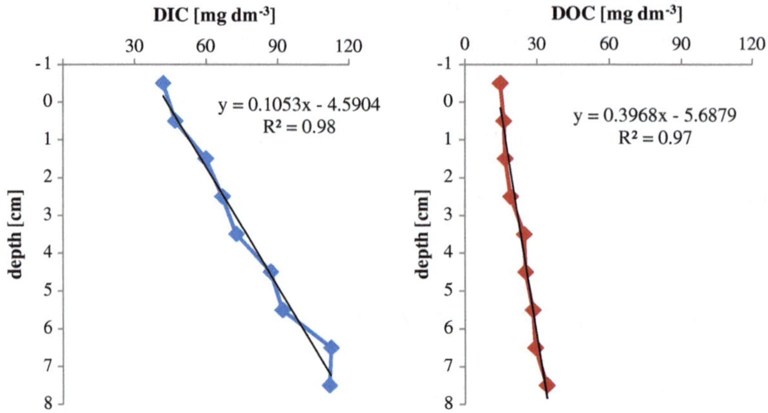

Fig. 4.8 The DIC and DOC concentration profiles in pore water from the upper layers of sediments of the Bornholm Deep. The determination coefficients (R^2) indicate the linear correlation strength. Accuracy of DIC and DOC concentrations analyses is described by relative standard deviation equal to 2.2% and 3.1%, respectively (n = 3) (modified after Kuliński and Pempkowiak 2011b)

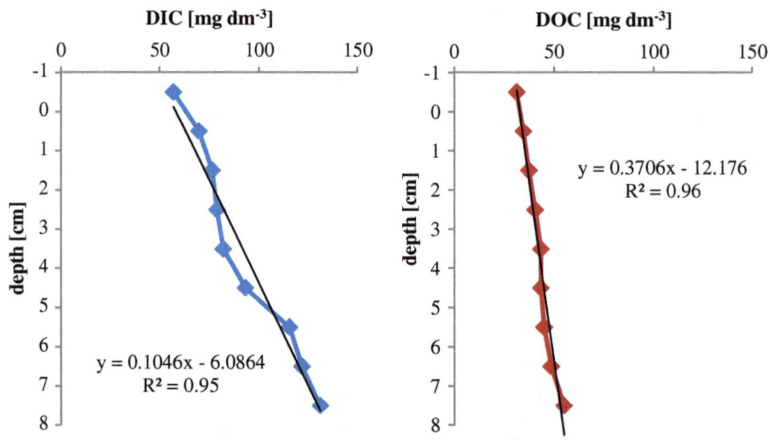

Fig. 4.9 The DIC and DOC concentration profiles in pore water from the upper layers of sediments of the Gdańsk Deep. The determination coefficients (R^2) indicate the linear correlation strength. Accuracy of DIC and DOC concentrations analyses is described by relative standard deviation equal to 2.2% and 2.8%, respectively (n = 3) (modified after Kuliński and Pempkowiak 2011b)

analyte concentration changes. Values of coefficient *a* tend to be lower for the DIC than for the DOC concentrations profiles, in agreement with higher values of the DIC gradients in pore water. The smallest DIC concentrations gradient is observed

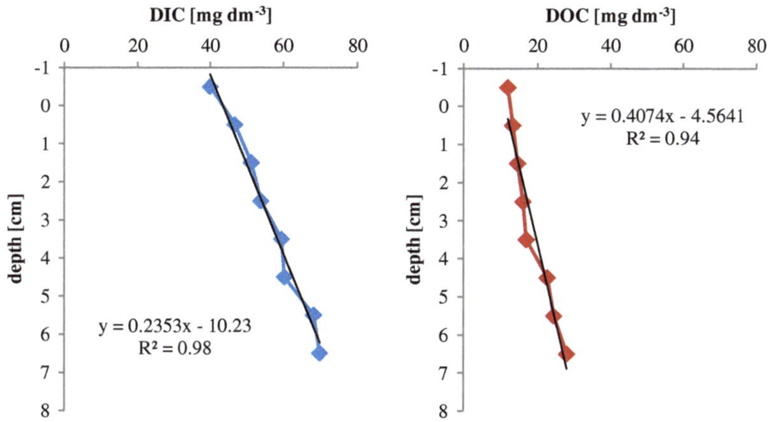

Fig. 4.10 DIC and DOC concentration profiles in pore water from the upper layers of sediments of the Gotland Deep. The determination coefficients (R^2) indicate the linear correlation strength. Accuracy of DIC and DOC concentrations analyses is described by relative standard deviation equal to 1.6% and 2.9%, respectively (n = 3) (modified after Kuliński and Pempkowiak 2011b)

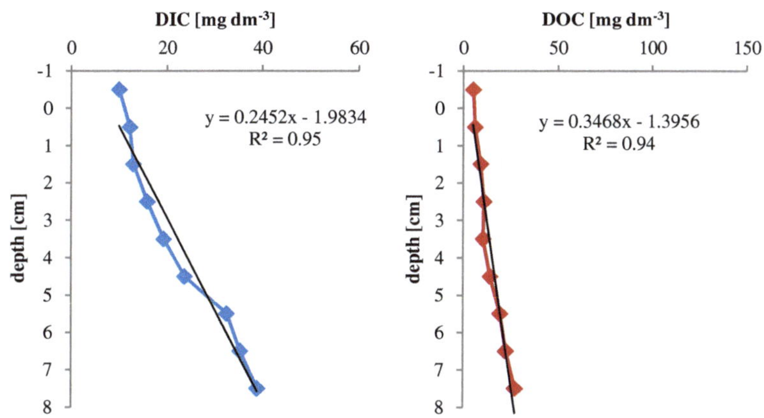

Fig. 4.11 The DIC and DOC concentration profiles in pore water from the upper layers of sediments of the Gulf of Bothnia. The determination coefficients (R^2) indicate the linear correlation strength. Accuracy of DIC and DOC concentrations analyses is described by relative standard deviation equal to 1.7% and 1.8%, respectively (n = 3) (modified after Kuliński and Pempkowiak 2011b)

for the sediment core from the Gulf of Bothnia (Fig. 4.11). Concentrations of DIC, in the investigated cores, oscillate from 10.2 ± 0.2 mg C dm^{-3} in the above-sediment water to 39.0 ± 0.7 mg C dm^{-3} at the 7–8 cm depth below sediment–water interface. In contrast, the Gdansk Deep DIC profile revealed the biggest DIC concentration changes (Fig. 4.9), with values ranging from 57.1 ± 1.2 mg C dm^{-3}

Table 4.6 Summary of diffusion magnitude and return fluxes of DIC and DOC from different depositional areas of the Baltic (Kuliński 2010)

Depositional area	Surface area (km^2)	C diffusion rate from sediments (g C m^{-2} $year^{-1}$)		C diffusion flux from sediments (Tg $year^{-1}$)		
		DIC	DOC	DIC	DOC	Total
Arkona Basin	3 200[a]	9.7	0.5	0.031	0.002	0.033
Bornholm Deep	13 170[a]	18.0	0.9	0.237	0.012	0.249
Gdansk Deep	12 150[a]	18.1	1.0	0.220	0.012	0.232
Gotland Deep	28 240[a,b]	8.6	0.9	0.243	0.025	0.268
Gulf of Riga	890[a]	8.6	0.9	0.008	0.001	0.009
Gulf of Finland	940[a]	8,6	0.9	0.008	0.001	0.009
Gulf of Bothnia	42 440[c]	7.0	1.0	0.297	0.042	0.339
Total	101,030	–	–	1.044	0.095	1.139

[a] (Błaszczyszyn 1982)
[b] (Emeis et al. 2000)
[c] (Algesten et al. 2006)

in the above-sediment water to 131.1 ± 2.9 mg C dm^{-3} in pore water collected at 7–8 cm.

In the case of DOC profiles, the highest *a* coefficient value was observed in pore water from the Arkona Deep (Fig. 4.7). Concentrations of DOC there vary between 14.7 ± 0.3 mg C dm^{-3} in the above-sediment water to 23.3 ± 0.5 mg C dm^{-3} in the layer at a depth of 6–7 cm. The biggest DOC concentrations gradient was observed in the pore water profile from the Gulf of Bothnia (Fig. 4.11), where DOC concentrations range between 5.3 ± 0.1 mg C dm^{-3} in the above-sediment water to 27.5 ± 0.5 mg C dm^{-3} at 7–8 cm depth.

Fluxes of dissolved carbon species, calculated based on the Fick's law of diffusion (see Appendix A.2 for details), are summarized in Table 4.6. The results indicate a strong return flux of carbon from sediments to the water column. Moreover, they point at DIC as the primary carrier of the carbon that returns from sediments to water, and, thus, at a dominance of mineralization processes over hydrolysis of organic matter deposited to the sediments. Diffusive fluxes of DIC range from 7.0 g C m^{-2} $year^{-1}$ in the Gulf of Bothnia to 18.1 g C m^{-2} $year^{-1}$ for the Gdansk Deep. Diffusive fluxes of DOC are an order of magnitude smaller and do not exceed 1.0 g C m^{-2} $year^{-1}$, as obtained for the Gdansk Deep and the Gulf of Bothnia. The minimum of 0.5 g C m^{-2} $year^{-1}$ is observed for the Arkona Basin Region.

The total carbon load released from sediments of the considered areas to the overlaying water amounts to 1.14 Tg C $year^{-1}$ (Table 4.6). The DOC flux contributes just 0.10 Tg C $year^{-1}$ ($<9\%$). The remaining 1.04 Tg C $year^{-1}$ reaches the water column as DIC. Depositional area characterized by the greatest absolute load of carbon is the Gulf of Bothnia, where the value amounts to 0.34 Tg C $year^{-1}$, which corresponds to nearly 30% of the total carbon released from the Baltic sediments to the above-sediments water. Regions characterized by the lowest contribution of the return flux are the Gulf of Riga and the Gulf of Finland. The contribution of both to the total carbon return flux from the sediments does not exceed 1%.

4.3.2 Labile Organic Matter Decomposition in Sediments

An independent method was used in order to confirm the magnitude of the carbon return flux from sediments. A sediment core collected in the Gdańsk Deep was stored for 12 months. In the course of the storage, organic carbon was measured in subsequent layers of the core at 4 month intervals. The core was dated by means of the ^{210}Pb method, and the modelled deposition ages were validated with ^{137}Cs distribution in the core (Pempkowiak 1991; Szczepańska et al. 2011).

Organic carbon concentration profiles in sediments after successive stages of incubation against deposition time are presented in Fig. 4.12. Data uncertainties for the subsequent series are expressed by the relative standard deviation (RSD) varying from 1.0% to 1.7% for samples of the second and the third incubation stages

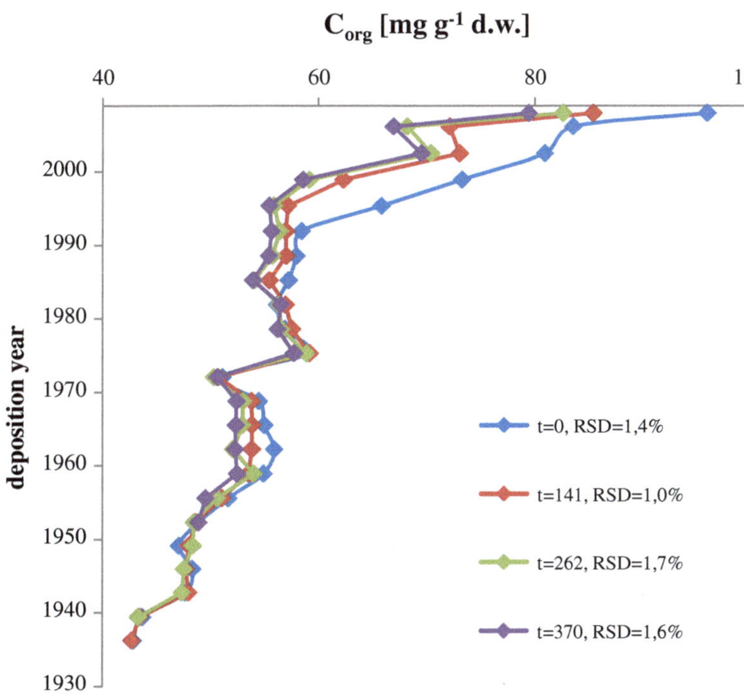

Fig. 4.12 Organic carbon concentration profiles in a Gdansk Deep core for the four incubation stages (Kuliński 2010)

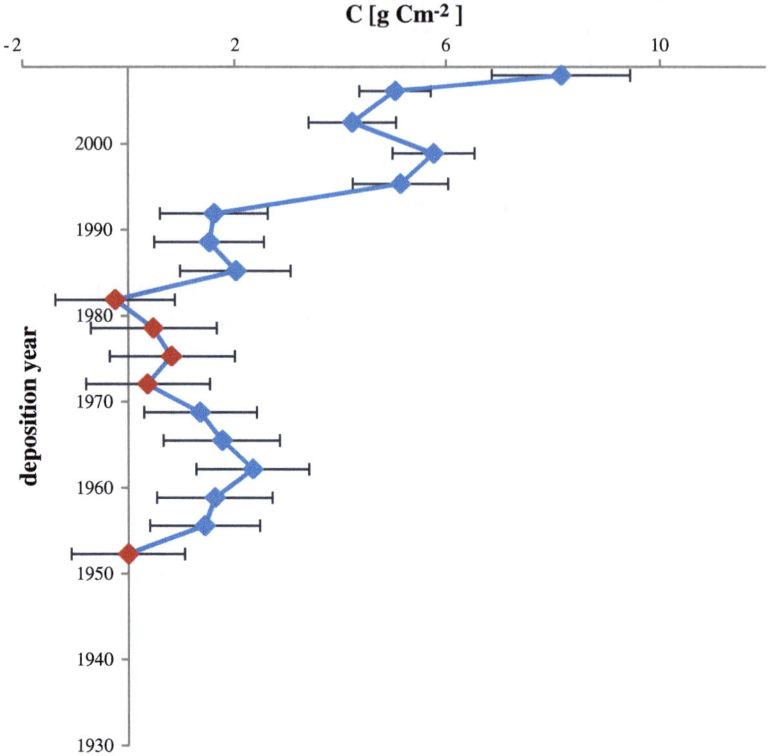

Fig. 4.13 Profile of the total organic carbon losses in the incubated sediment core from the Gdansk Deep. Red squares indicate sediment layers characterized by negative values of the lower uncertainty intervals (Kuliński 2010)

respectively. The highest concentrations of organic carbon in the analyzed sediment cores are found in samples analyzed at the beginning of the incubation time (t = 0), with values ranging from 42.9 ± 0.6 mg C g^{-1} d.w. for the layer deposited in 1936 to 96.0 ± 1.34 mg C g^{-1} d.w. for semi-liquid, topmost layer ("fluffy layer"), deposited in 2008. The lowest values were observed for core incubated for 370 days and varied from 49.0 ± 0.8 mg C g^{-1} d.w. in the layer dated on 1952 to 79.4 ± 1.3 mg C g^{-1} d.w. in the "fluffy layer". All four profiles are characterized by a noticeable concentration increase during the early 1990s, when also the biggest differences between subsequent layers of the sediment core are recorded. In sediments deposited before 1990, concentration values are more comparable, and the decrease with depth is less pronounced. For all analyzed cores, a characteristic drop in the organic carbon concentration occurs in the layer dated on 1972. This situation could be a result of sediments re-suspension and re-deposition event in the early seventies. The lack of differences between the organic carbon concentrations at successive stages of incubation indicate that the sediments in the layer are actually older than suggested by ^{210}Pb dating method. Below this level, up to a layer dated on 1956, a slightly higher organic carbon concentrations are observed, as well as distinct differences in concentrations for the subsequent stages of incubation are recorded. The deeper layers are characterized by lower carbon concentrations, lacking of statistically significant differences for successive stages of incubation, which indicate organic matter mineralization and decomposition processes to be completed or nearly completed.

Organic carbon total losses (C_{min}) from successive 1 cm layers of the analyzed sediment core are expressed as g C m^{-2}, as presented in Fig. 4.13. Values are extrapolated to the temperature of 4°C. For the entire incubation experiment, the most significant losses in organic carbon occurred for layers deposited after 1992, when C_{min} ranges from 4.2 ± 0.8 g C m^{-2} for the layer deposited in 2003 to 8.2 ± 1.3 g C m^{-2} for the freshly deposited "fluffy layer". Significant organic carbon losses in this part of the core indicate intensive organic matter mineralization and decomposition taking place during a time scale of less than 15 years. Below this depth in the core, carbon losses from the sediments are still occurring, but the scale of the process is less significant than in the upper level and does not exceed 2.3 ± 1.0 g C m^{-2} in a layer dated on 1962. Despite the fact that the analyzed data encompass the period until 1952, due to a shorter time span characterizing the fourth phase of incubation, it can be assumed that in the deeper layers the organic matter mineralization and decomposition processes are no longer present. The explanation comes from a lack of statistically significant differences between the organic carbon concentration differences, in the considered core fragment, during earlier phases of incubation (Fig. 4.12). Figure 4.13 emphasizes four layers (marked in red) for which negative, or close to negative values of carbon losses were observed. Such sediments are characterized by organic matter that can be regarded as chemically stable under the storage conditions, as well as the absence or close-to-minimum rate of mineralization and decomposition processes. Based on the results obtained, the total organic carbon loss was estimated to be 35.5 ± 11.8 g C m^{-2}. The loss can be attributed to the mineralization of labile, and semi-stable fractions of organic matter spread along the core.

The results of organic carbon loss rates (v) for the sediment core are based on the first degree kinetic equation that represents the organic matter mineralization velocity (Fig. 4.14). The fastest rates are observed during the first phase of incubation, especially in the upper sediment layer, deposited after 1992. Values range here from 22.1 mg C m^{-3} day^{-1} in a layer dated for 2003 to 39.5 mg C m^{-3} day^{-1} for the "fluffy layer". In the course of the further stages of the experiment the observed rates of the organic matter losses are slower within the upper segment of the core, that is, in sediments deposited in the period from 1995 to 2008. The obtained results vary in the range from 1.9 mg C m^{-3} day^{-1} for sediments deposited in 1995 to 15.2 mg C m^{-3} day^{-1} for the "fluffy layer"—both recorded during the third phase of incubation. An average rate calculated for the entire

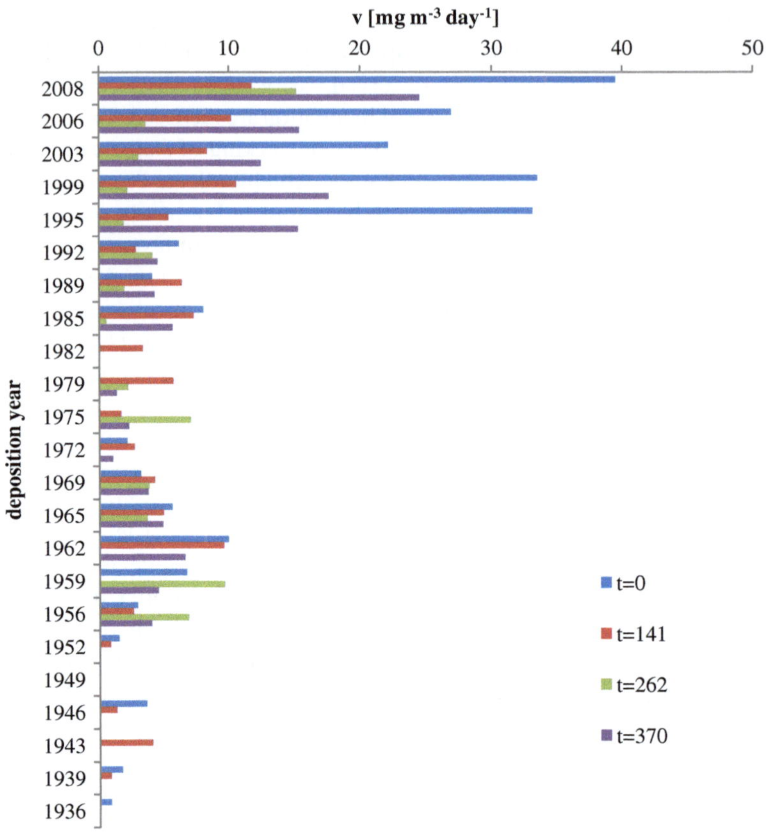

Fig. 4.14 Daily rates of the organic carbon loss during the successive phases of the sediment core incubation (Kuliński 2010)

experiment is driven by its high values observed for the first phase, and the obtained results fall between 12.5 mg C m^{-3} day^{-1} for the sediment deposited in 2003 to 24.5 mg C m^{-3} day^{-1} for the "fluffy layer". Below the uppermost sediment layers, the daily rates of the organic carbon loss are smaller, not exceeding 9.9 mg C m^{-3} day^{-1}, estimated during the first phase of incubation, for the material deposited in 1962. Slight differences are found for specific phases of the experiment, as well as for the average of the entire period during which the experiment was run. A pronounced increase in the organic matter decomposition rates, similar to those for organic carbon concentrations and C_{min} (Fig. 4.13), is observed in layers deposited between 1956 and 1969. However, differences in the organic carbon loss rates at successive stages of the experiment are minor for this core segment, in contrast to the uppermost layer of sediments. A different situation is observed in the

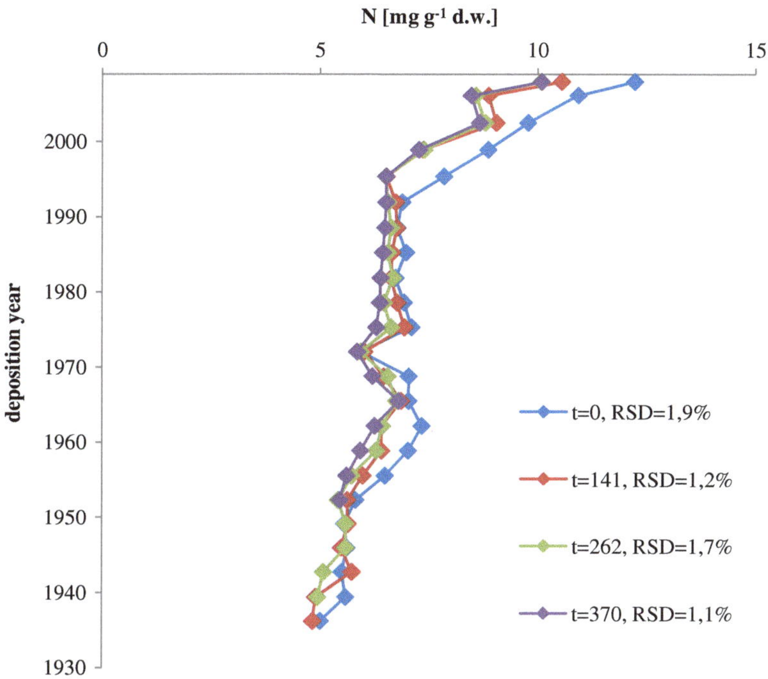

Fig. 4.15 Total nitrogen profile obtained for successive phases of the Gdansk Deep sediment core incubation (Kuliński 2010)

uppermost layer of the core, where the rates of mineralization during the second and the third phase of incubation are three times higher than those during the first phase. The results obtained in the experiment indicate the presence of semi-stabile organic matter whose contribution gradually decreases with depth and, at the same time, with the sediment age; its half-time is estimated at to be 55–60 years. Within the first 15 years the mineralization rate is significantly more intensive.

Profiles of the total nitrogen and phosphorus in the sediment cores subjected to the incubation (Figs. 4.15, 4.16) are in close agreement with the organic carbon concentrations profiles (Fig. 4.12). Relative standard deviations (RSD) characterizing the successive measurement series vary from 1.1 to 1.9% and from 1.6 to 2.2% for nitrogen and phosphorus, respectively. In the uppermost sediment layers deposited after 1992, the concentration of N and P increases, reaching its maximum in the most recently deposited sediment layers, amounting to 12.2 ± 0.2 mg N g^{-1} d.w. and 0.90 ± 0.02 mg P g^{-1} d.w. Similarly to the organic carbon profiles, N and P concentrations in a layer dated on 1972 are characterized by a distinct decrease, reaching 5.9 ± 0.1 mg N g^{-1} and

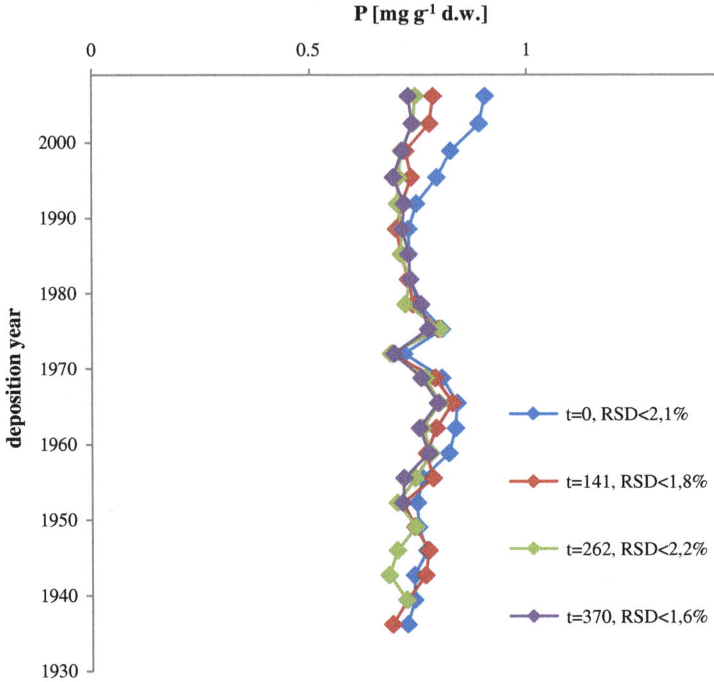

Fig. 4.16 Total phosphorus profile obtained for successive phases of the Gdansk Deep sediment core incubation (Kuliński 2010)

0.72 ± 0.2 mg P g^{-1}, whereas for the subsurface sediment layers, deposited between 1956 and 1969, a slight increase of N and P concentrations is observed, as well as an increase in concentration amplitude between subsequent incubation stages.

An analysis of C/N molar ratios (Table 4.7) reveals rather steady values in vertical profiles that range from 8.9 in layers dated on 1962 and 2006 to 10.1 in layers dated on 1943 and 1972, with an average of 9.6. A small increase is observed during the course of incubation for the subsequent sediment layers. The end of the experiment is characterized by values that oscillate in the range from 9.2 for the "fluffy layer" to 10.5 for the layer dated on 1952. Another pattern represents the C/P molar ratios profile, exceeding even 110 for the cores upper layer, and a value smaller than 80 in the deepest layers. Similar 'depth' trend is characteristic of the N/P molar ratios. In both cases, C/P and N/P profiles do not reveal any systematic changes with time.

Table 4.7 Summary of C/N, C/P and N/P molar ratios at different incubation stages of the Gdansk Deep sediment core (Kuliński 2010)

Deposition year	Incubation start			1st step of incubation			2nd step of incubation			3rd step of incubation		
	C/N	C/P	N/P	C/N	C/P	N/P	C/N	C/P	N/P	C/N	C/P	N/P
2008	9.2	–	–	9.5	–	–	9.6	–	–	9.2	–	–
2006	8.9	115.5	13.0	9.5	114.6	12.1	9.3	114.4	12.3	9.2	114.8	12.4
2003	9.7	113.4	11.7	9.4	117.3	12.5	9.3	118.8	12.7	9.4	117.8	12.6
1999	9.7	110.8	11.5	9.9	107.8	10.9	9.4	103.9	11.1	9.4	102.5	10.9
1995	9.8	103.5	10.6	10.2	97.2	9.5	10.0	99.0	9.9	9.9	100.1	10.1
1992	9.9	97.8	9.9	9.9	99.9	10.1	10.0	100.4	10.0	10.0	96.9	9.7
1989	10.0	99.4	9.9	9.9	101.8	10.3	9.8	97.8	9.9	10.0	97.0	9.7
1985	9.6	98.1	10.2	9.7	97.4	10.0	9.6	95.1	9.9	9.8	92.6	9.5
1982	9.7	95.9	9.8	10.0	97.9	9.8	9.9	96.3	9.8	10.3	96.5	9.3
1979	9.6	94.6	9.8	9.9	97.3	9.8	10.2	98.0	9.6	10.3	92.8	9.0
1975	9.7	91.6	9.4	10.0	92.8	9.3	10.4	91.8	8.8	10.7	93.3	8.7
1972	10.1	89.1	8.8	9.9	92.0	9.3	9.9	91.6	9.2	10.1	91.2	9.0
1969	9.1	84.7	9.4	9.7	85.3	8.8	9.5	86.5	9.1	9.9	86.6	8.8
1965	9.1	81.9	9.0	9.2	81.2	8.9	9.2	82.9	9.0	9.0	82.3	9.1
1962	8.9	83.6	9.4	9.8	85.0	8.7	9.5	85.7	9.0	9.8	86.9	8.9
1959	9.1	83.7	9.2	9.8	87.1	8.9	10.0	86.7	8.6	10.4	84.7	8.2
1956	9.3	85.2	9.2	10.0	81.5	8.2	10.3	85.2	8.2	10.3	86.5	8.4
1952	9.8	81.9	8.3	10.1	85.2	8.4	10.5	86.5	8.3	10.5	85.7	8.2
1949	9.9	78.6	7.9	10.0	80.9	8.1	10.1	81.4	8.0	–	–	–
1946	10.0	78.6	7.8	10.2	77.2	7.6	9.9	84.8	8.5	–	–	–
1943	10.1	80.6	8.0	9.8	78.3	8.0	10.9	86.9	8.0	–	–	–
1939	9.1	73.8	8.1	10.3	75.0	7.3	10.2	74.9	7.3	–	–	–
1936	10.0	73.9	7.4	10.3	77.3	7.5	–	–	–	–	–	–

4.4 Carbon Inflow to the Baltic with Precipitation

To illustrate the magnitude of carbon inflows and losses, Algesten et al. (2006) present comprehensive approach of organic carbon loads transported by different atmospheric deposition forms. With annual precipitation equal to 237 km^3 (Lass and Matthäus 2008), the total organic carbon reaching the Baltic Sea has been estimated at 0.45 Tg C year^{-1}. Aerosols are considered to be the main source of carbon in rainwater (Abril et al. 2002). For samples collected above the Baltic open water area, the organic carbon concentration has been estimated to be eight times higher than those recorded for the particulate carbon present mainly as soot (Kuśmierczyk-Michulec et al. 2001). Based on this conclusions, a similar ratio has been assumed for the rainwater, which gives a positive load of 0.06 Tg C reaching the Baltic each year.

For the purpose of assessing an inorganic carbon load carried with the rainwater, the aerosol fraction has not been included in calculations. It has been also assumed that the aerosols emitted from the Baltic surface are compensated through the dry and wet re-deposition.

The atmospheric CO_2 dissolved in rainwater has been recognized as the Baltic Sea inorganic carbon source. The CO_2 dissolved in rainwater is assumed to be in equilibrium with the atmospheric CO_2 characterized by partial pressure of 385 ppm (value for 2009, NOAA records, www.esrl.noaa.gov).

Using Henry's law (Eq. 4.1) for a temperature of 10°C and atmospheric pressure of 1013 hPa, the CO_2 concentration in rainwater was calculated at 0.24 mg C dm^{-3} (Ibanez et al. 2007). When multiplied by the annual precipitation (237 km^3), the annual carbon load to the Baltic equal to 0.06 Tg C $year^{-1}$ is obtained. Henry's law equation reads:

$$C = K_H \cdot p \tag{4.1}$$

where: C—CO_2 concentration in a rainwater (mol dm^{-3})

K_H—Henry's law constant (mol dm^{-3} atm^{-1})
p—CO_2 partial pressure in the atmosphere (atm)

It is difficult to assess the annual load of organic carbon brought to the Baltic with precipitation, since no data regarding the DOC and/or POC concentration in the rainwater have been reported. Taking the value of TOC in the rainwater reported by Pempkowiak (1976), the annual load of organic carbon deposited to the Baltic amounts to 0.24 Tg C $year^{-1}$. This approximation should be taken with caution, since the assumed concentrations have not been confirmed by recent reports. Nevertheless, the figure approximates the magnitude of organic carbon delivered to the Baltic with rainwater. Both DIC and TOC loads deposited to the Baltic with rainwater are negligible as compared to other sinks and sources of carbon to the Baltic.

References

Abril G, Nogueira M, Etcheber H, Cabeçadas G, Lemaire E, Brogueira MJ (2002) Behaviour of organic carbon in nine contrasting european estuaries. Estuarine, Coast Shelf Sci 54:241–262

Algesten G, Brydsten L, Jonsson P, Kortelainen P, Löfgren S, Rahm L, Räike A, Sobek S, Tranvik L, Wikner J, Jansson M (2006) Organic carbon budget for the Gulf of Bothnia. J Marine Syst 63:155–161

Błaszczyszyn A (1982) Główne składniki chemiczne osadów dennych Morza Bałtyckiego (Main chemical components of bottom sediments in the Baltic Sea), In: Gudelis WK, Jemielianow JM (eds) Geologia Morza Bałtyckiego. Wydawnictwa Geologiczne, Warszawa, pp 257–289 (in Polish)

Christoffersen PL, Christiansen C, Jensen JB, Leipe T, Hille S (2007) Depositional conditions and organic matter distribution in the Bornholm Basin, Baltic Sea. Geo-Marine Lett 27:325–338

Emeis K-C, Struck U, Leipe T, Pollehne F, Kunzendorf H, Christiansen C (2000) Changes in the C, N, P burial rates in some Baltic Sea sediments over the last 150 years–relevance to P regeneration rates and the phosphorus cycle. Marine Geol 167:43–59

Ibanez JG, Hernandez-Esparza M, Doria-Serrano C, Fregoso-Infante A, Singh MM (2007) Environmental Chemistry. Springer Science + Business Media, LLC, New York, p 325

Kuliński K (2010) Obieg węgla w Morzu Bałtyckim (Carbon cycling in the Baltic Sea). PhD thesis, Institute of Oceanology PAS, p 134 (in Polish)

Kuliński K, Pempkowiak J (2011a) The carbon budget of the Baltic Sea. Biogeosciences Discuss. 8:4841–4869. doi:10.5194/bgd-8-3219-2011

Kuliński K, Pempkowiak J (2011b) Accumulation, mineralization and burial rates of organic carbon in the Baltic Sea sediments. Marine Chem (submitted)

Kuliński K, She J, Pempkowiak J (2011) Short and medium term dynamics of the carbon exchange between the Baltic Sea and the North Sea. Cont Schelf Res 31:1611–1619

Lass H-U, Matthäus W (2008) General Oceanography of the Baltic Sea. In: Feistel R, Nausch G, Wasmund N (eds) State and evolution of the Baltic Sea, 1952–2005. Wiley, Hoboken, NJ, pp 5–43

Pempkowiak J (1991) Enrichment factors of heavy metals in the Southern Baltic surface sediments dated with 210Pb and 137Cs. Environ Int 17:421–428

Prowe AEF, Thomas H, Pätsch J, Kühn W, Bozec Y, Schiettecatte L-S, Borges AV, de Baar HJW (2009) Mechanisms controlling the air-sea CO_2 flux in the North Sea. Cont Shelf Res 29:1801–1808

Suratman S, Weston K, Jickells T, Fernand L (2009) Spatial and seasonal changes of dissolved and particulate organic C in the North Sea. Hydrobiol 628:13–25

Szczepańska A, Maciejewska A, Zaborska A, Kuliński K, Pempkowiak J (2011) Distribution and provenience of organic carbon in the Baltic sediments. Geochronometria (accepted)

Thomas H, Schneider B (1999) The seasonal cycle of carbon dioxide in Baltic Sea surface waters. J Marine Syst 22:53–67

Thomas H, Bozec Y, de Baar HJW, Elkalay K, Frankignoulle M, Schiettecatte L-S, Kattner G, Borges AV (2005) The carbon budget of the North Sea. Biogeosciences 2:87–96

Chapter 5
Research on Carbon Cycling in the Baltic: Discussion

5.1 The Baltic Sea Carbon Budget

The carbon fluxes quantified in Chap. 4 give a complete outlook of the carbon sources (positive values) and sinks (negative values) in the Baltic Sea (Fig. 5.1). The Baltic carbon cycle is characterized by a balance between carbon inflows and outflows to/from the basin (see Eq. 5.1). Thus, quantifying carbon fluxes and substituting appropriate values in Eq. 5.2, one gets the total carbon inflow exceeding the total outflow by 1.14 Tg C year^{-1}. The obtained difference is associated with the CO_2 net exchange between the seawater and the atmosphere (Eq. 5.3). Therefore, the estimates indicate the Baltic Sea as a net source of CO_2 to the atmosphere amounting to -1.14 Tg C year^{-1}, which corresponds to -4.18 Tg CO_2 year^{-1} (Fig. 5.1). The equations quoted above read:

$$\sum \text{carbon sources} = \sum \text{carbon sinks} \tag{5.1}$$

$$F_e + F_i + F_o + F_a + F_f + F_p + F_r + F_m + F_s = 0 \tag{5.2}$$

$$-F_a = F_e + F_i + F_o + F_f + F_p + F_r + F_m + F_s \tag{5.3}$$

where:

F_e export to the North Sea
F_i import from the North Sea
F_o precipitation
F_a CO_2 net exchange between water and atmosphere
F_f fisheries
F_p point sources
F_r rivers
F_m sediments return flux
F_s deposition to sediments

K. Kuliński and J. Pempkowiak, *Carbon Cycling in the Baltic Sea*,
Geoplanet: Earth and Planetary Sciences, DOI: 10.1007/978-3-642-19388-0_5,

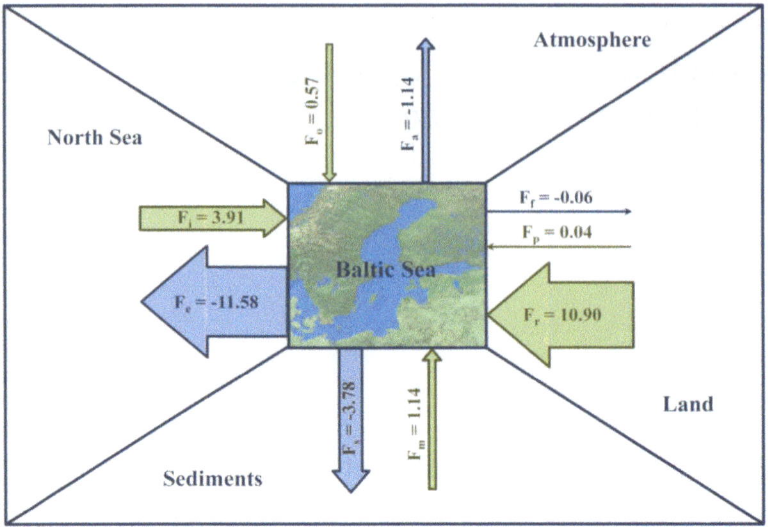

Fig. 5.1 A scheme of the carbon budget in the Baltic Sea. Carbon inflows in green, outflows in blue. The arrow width is directly proportional to the carbon flux. Carbon fluxes are expressed in [Tg C year^{-1}]

Assuming constant CO_2 net emission from each unit area of the basin surface and the total Baltic area of 3.85 10^5 km^2 (Kattegat excluded) (Gudelis and Jemielianow 1982; BACC Author Team 2008), the average CO_2 net flux (F_a) to the atmosphere equals -3.0 gC m^{-2} year^{-1}, which corresponds to -11.0 g m^{-2} year^{-1} of CO_2.

The obtained value suggests that the Baltic Sea is characterized by a moderately neutral balance of the CO_2 exchange between water and the atmosphere. Also, results of mathematical modelling by Omstedt et al. (2009) indicate the basin as CO_2 source, or the atmospheric CO_2 absorption, area depending on the season. Among the available literature data concerning CO_2 exchange between the atmosphere and different regions of the Baltic Sea (Table 5.1), only those presented by Algesten et al. (2004, 2006) for the Gulf of Bothnia reveal CO_2 net emission to the atmosphere. Both estimates are quite close: -37.2 gC m^{-2} year^{-1} (Algesten et al. 2004) compared to -35.4 g C m^{-2} year^{-1} (Algesten et al. 2006). Results differing from those by Algesten et al. (2006) were obtained for other Baltic Sea regions. For the so-called Baltic Proper, together with the Gulf of Finland and the Gulf of Riga, two separate estimates indicate a net absorption of the atmospheric CO_2 by the seawater equal to 10.8 g C m^{-2} year^{-1} (Ohlson 1990) and 13.2 g C m^{-2} year^{-1} (Thomas and Schneider 1999). Regular studies conducted in the Arkona Basin (Kuss et al. 2006) indicate even greater uptake, reaching 36.0 g C m^{-2} year^{-1}.

Table 5.1 Summary of the average CO_2 emission/absorption values for different regions of the basin

Region	F_a (g C m^{-2} $year^{-1}$)	Data source
Baltic Sea	−3.0	this book
Baltic Sea	2.3	Thomas et al. (2010)
Baltic Proper + G.Finland + G. Riga	13.2	Ohlson (1990)
Baltic Proper + G.Finland + G. Riga	10.8	Thomas and Schneider (1999)
Arkona Basin	36.0	Kuss et al. (2006)
Gulf of Bothnia	−37.2	Algesten et al. (2004)
Gulf of Bothnia	−35.4	Algesten et al. (2006)

Negative values indicate CO_2 emission from the sea surface to the atmosphere, positive values indicate the atmospheric CO_2 absorbed by seawater

In the situation when the total CO_2 net load emitted by the Baltic Sea, as determined in the study presented here (−1.14 Tg C $year^{-1}$), is reduced by the CO_2 flux of the Gulf of Bothnia (−3.61 Tg C $year^{-1}$), one obtains the amount of carbon absorbed by the remaining regions of the Baltic Sea, i.e., the Baltic Proper with the Gulf of Finland and the Gulf of Riga. The result is the net absorption of 2.47 Tg C $year^{-1}$ in the form of CO_2. Assuming the total area of the Baltic Proper, the Gulf of Finland and the Gulf of Riga to be 2.83 10^5 km^2 (Gudelis and Jemielianow 1982; Algesten et al. 2006; BACC Author Team 2008), the average atmospheric CO_2 flux to the sea is 8.7 g C m^{-2} $year^{-1}$, which is smaller by 2.1 g C m^{-2} $year^{-1}$ (some 20%) than the value reported by Thomas and Schneider (1999) and more than 30% smaller than value presented by Ohlson (1990) (Table 5.1). The absolute values of CO_2 emission/absorption results obtained for the Baltic Sea are smaller than those observed for other shelf seas (Table 5.2). Among the basins considered, the Arabian Sea (−10 g C m^{-2} $year^{-1}$; Goyet et al. 1998) and single reports for the North Sea (−9.4 g C m^{-2} $year^{-1}$; Prowe et al. 2009), the Bering Sea (−56.4 g C m^{-2} $year^{-1}$; Fransson et al. 2006) and South China Sea (−15.6 g C m^{-2} $year^{-1}$; Zhai et al. 2005) indicate the CO_2 emission to the atmosphere. Comparable values of carbon dioxide absorption to those observed for the Baltic and its adjacent gulfs are found for the Sea of Okhotsk (10 g C m^{-2} $year^{-1}$; Wakita et al. 2003) and the Baffin Bay (5.2 g C m^{-2} $year^{-1}$; Miller et al. 2002), as well as for selected basins of the Barents Sea (6.6 g C m^{-2} $year^{-1}$; Fransson et al. 2001) and the South China Sea (12.0 g C m^{-2} $year^{-1}$; Chen et al. 2003).

Among the shelf seas for which CO_2 emissions/absorptions are summarized in Table 5.2, those localized within the subpolar zone of the Pacific Ocean absorb the atmospheric CO_2 in the most intensive manner. The Chukchi Sea has been estimated to absorb 57.6 g C m^{-2} $year^{-1}$ (Bates 2006) and the Bering Sea—51.6 g C m^{-2} $year^{-1}$ (Walsh and Dieterle 1994). Interestingly, the highest CO_2 emission values (Table 5.2) are also found for the Bering Sea, especially within its upwelling zones, reaching as much as −56.4 g C m^{-2} $year^{-1}$ (Fransson et al. 2006). The results presented in Table 5.2 reveal, perhaps, significant data discrepancy within the individual basins, which could be caused by specific locations

Table 5.2 Summary of the average CO_2 emission/absorption (FCO_2) for the Baltic Sea, and other shelf seas of the northern hemisphere

Region	FCO_2 (g C m^{-2} $year^{-1}$)	Data source
Baltic Sea	−3.0	this book
Baltic Proper + G.Finland + G.Riga	8.7	Kuliński (2010) based on Algesten et al. (2006)
Gulf of Bothnia	−35.4	Algesten et al. (2006)
North Sea	18.0–26.4	Bozec et al. (2005)
North Sea	16.5	Thomas et al. (2005)
North Sea–southern part	−9.4	Prowe et al. (2009)
North Sea–northern part	24.7	Prowe et al. (2009)
Bering Sea	6.6	Fransson et al. (2001)
Bering Sea	43.2	Borges et al. (2005)
Bering Sea	51.6	Walsh and Dieterle (1994)
Bering Sea–upwelling zone	−56.4	Fransson et al. (2006)
Chukchi Sea	57.6	Bates (2006)
Chukchi Sea	37.2	Kaltin and Anderson (2005)
Sea of Okhotsk	10.0	Wakita et al. (2003)
Baffin Bay	5.2	Miller et al. (2002)
Arabian Sea	−10.8	Goyet et al. (1998)
Sea of Japan	45.6	Cai et al. (2006)
South China Sea	−15.6	Zhai et al. (2005)
South China Sea	12.0	Chen et al. (2003)
Yellow Sea	24.0	Chen and Borges (2009)

of the research stations. This situation indicates a significant variation as regards intensity and the exchange direction of CO_2 between shelf seas and the atmosphere. The Baltic Sea also falls into this category, with an average CO_2 emission to the atmosphere from the Gulf of Bothnia to be −35.4 g C m^{-2} $year^{-1}$ (Algesten et al. 2006), compared to an average absorption by the Baltic Proper and its adjacent gulfs (Gulf of Riga and Gulf of Finland) to be 8.7 g C m^{-2} $year^{-1}$. The situation described above is another reason to avoid assumptions and generalizations regarding the role of the seas as sources/sinks of carbon dioxide. The discrepancies for different regions of shelf seas, including Baltic, can be explained by emission/absorption results obtained for estuaries (Table 5.3). Within an estuary, well-oxygenated seawater meets river water. The latter carries, most often, significant amounts of organic matter, labile organic substances included. These on mixing with sea water, are subjected to intensive mineralization. Moreover, river water carries inorganic carbon species, including dissolved CO_2. This results in oversaturation of the estuarine water as regards carbon dioxide and degassing to the atmosphere significant quantities of CO_2. Emission rates recorded for the estuaries exceed greatly those observed within upwelling zones, where CO_2 is released to the atmosphere, after it had been cumulated in subsurface water masses for decades, as an effect of organic matter mineralization (Fransson et al. 2006).

Table 5.3 Selected literature data concerning CO_2 emissions to the atmosphere from estuaries

Region	FCO_2 (g C m^{-2} $year^{-1}$)	Literature source
Rhine estuary	−476.4	Frankignoulle et al. (1998)
Elbe estuary	−636.0	Frankignoulle et al. (1998)
Thames estuary	−883.2	Frankignoulle et al. (1998)
Scheldt estuary	−756.0	Frankignoulle et al. (1998)
Loire estuary	−772.8	Abril et al. (2003)

Among the estuaries listed in Table 5.3, CO_2 emission ranges from 476.4 g C m^{-2} $year^{-1}$ for the Rhine, to 883.2 g C m^{-2} $year^{-1}$ for the Thames (Frankignoulle et al. 1998).

A river influence has been well recognized and visible for open water regions. Based on results obtained at two independent stations localized within the northern and southern parts of the North Sea, Prowe et al. (2009) studied the magnitude and direction of CO_2 exchange between the atmosphere and seawater. The entire North Sea, with its northern and central parts in particular, is under a strong influence of the Atlantic Ocean. The feature that differentiates the southern part of the basin relays on a significant river runoff. Significant quantities of organic matter are carried with riwerwater, e.g., by Rhine, Scheldt, and Thames (Frankignoulle et al. 1998; Prowe et al. 2009). This is reflected in the direction and average annual values of the net CO_2 exchange between seawater and the atmosphere. The "northern station" revealed CO_2 absorption equal to 24.7 g C m^{-2} $year^{-1}$, while the "southern station" waters demonstrated CO_2 emission of −9.4 g C m^{-2} $year^{-1}$ (Prowe et al. 2009).

The above facts suggest that the emission/absorption differences for the Baltic Sea sub-regions (the Gulf of Bothnia versus Baltic Proper with the Gulf of Finland and the Gulf of Riga) are, in fact, significantly greater than the average values presented above. Thus, for estuarine regions, intensive CO_2 degassing to the atmosphere is anticipated. In the southern Baltic such processes are likely to occur in areas of main river estuaries: Neman River, Vistula, Odra. In the Gulf of Riga and the Gulf Finland, where estuaries of the Neva and the Daugava are localized, the situation is comparable. On the other hand, in order for the average of 8.7 g C m^{-2} $year^{-1}$ to be representative for the entire Baltic Proper with adjacent bays, the open basins have to intensively absorb the atmospheric CO_2. This hypothesis has been confirmed by Kuss et al. (2006), who estimated the atmospheric CO_2 absorption for the Arkona Basin at 36.0 g C m^{-2} $year^{-1}$ (Table 5.1). For the highly eutrophized southern Baltic (Nausch et al. 2008; HELCOM 2009), the biological pump becomes the determining factor for the atmospheric CO_2 absorption. Intensive phytoplankton blooms, stimulated by significant nutrient supplies (HELCOM 2009), contribute to enhanced atmospheric CO_2 absorption (Thomas and Schneider 1999; Kuss et al. 2006) that can be expressed by primary productivity reaching as much as 200 g C m^{-2} $year^{-1}$ (Wasmund et al. 2001; Wasmund and Siegel 2008). The magnitude of the primary productivity—the driving force for the atmospheric CO_2 absorption—varies for the individual Baltic

Sea regions. Its lowest values have been found for the Gulf of Bothnia, where in its northern part the primary productivity is as low as 20 g C m^{-2} $year^{-1}$ (Wasmund et al. 2001). This area is characterized by phosphorus deficit conditions (known to control the phytoplankton growth), as well as less favourable climatic and light conditions that shorten the vegetation period, as compared to the Baltic Proper (Andersson et al. 1996; Wasmund and Siegel 2008). The biological pump intensity is, at the same time, definitely lower than in the Gulf of Bothnia. Since the river runoff carry significant organic carbon loads to the Gulf of Bothnia (Table 4.4), it makes the basin an important source of CO_2 to the atmosphere (Algesten et al. 2004 and 2006). Simultaneously, the entire Baltic becomes an area characterized by a moderate CO_2 emission to the atmosphere (−3.0 g C m^{-2} $year^{-1}$ on average). However, it should be kept in mind that the southern Baltic is a strong sink of the carbon dioxide.

5.2 Factors Determining Carbon Budget in the Baltic Sea

Considerations presented in Sect. 5.1 indicate the presence of diverse trophic states within the individual Baltic regions. Strong absorption of the atmospheric CO_2 by the Baltic Proper or the Arkona Basin suggests their autotrophic character. The amount of CO_2 assimilated during photosynthesis exceeds here that released during the respiration. On the other hand, estuaries and, on average, the entire Gulf of Bothnia and the whole of the Baltic Sea, which are the areas emitting CO_2 to the atmosphere, are heterotrophic since a net loss of organic carbon is observed there (Table 5.4).

Quantitative budget of carbon sources and losses encountered in the Baltic Sea (Fig. 5.1; Table 5.4) clearly identifies inorganic carbon (IC) as prevailing over organic carbon (OC). The three most important fluxes that shape carbon budget in the basin, and thus CO_2 exchange between seawater and the atmosphere, are: carbon inflow with river runoff, exchange with the North Sea, and organic matter deposition/degradation within sediments.

The largest source of both inorganic and organic carbon species are rivers escaping to the Baltic. They are responsible for 66% (10.90 Tg C $year^{-1}$) of the total carbon load discharged to the basin (16.56 Tg C $year^{-1}$), while the share is even more significant when organic carbon (83%) is considered.

The second largest IC source to the Baltic is the inflow from the North Sea (3.70 Tg C $year^{-1}$). The OC fraction imported from the North Sea plays a less important role (0.21 Tg C $year^{-1}$)—somewhat above 5% of the total carbon imported to the basin. The OC fraction plays a more important role in the flux escaping the Baltic through the Danish Straits, where it accounts for 16% (−1.88 Tg C $year^{-1}$) compared to 84% (−9.70 Tg C $year^{-1}$) of IC fraction contribution. At the same time, carbon exchange with the North Sea becomes the most important rout of carbon export from the Baltic Sea, amounting to a net value of −7.67 Tg C $year^{-1}$, a load smaller by 43% than the estimate presented by Thomas

Table 5.4 Summary of carbon sources and losses in the Baltic (Tg C year^{-1})

Flux	Sources		Losses		Sum		Total
	IC	OC	IC	OC	IC	OC	
Rivers	6.81	4.09			6.81	4.09	10.90
North Sea	3.70	0.21	−9.70	−1.88	−6.00	−1.67	−7.67
Sediments	1.04	0.10		−3.78	1.04	−3.68	−2.64
Precipitation	0.06	0.51[a]			0.06	0.51[a]	0.57
Wastewater		0.04				0.04	0.04
Fisheries				−0.06		−0.06	−0.06
CO_2 Net exchange with the atmosphere			−1.14		−1.14		−1.14
Sum	11.61	4.95	−10.84	−5.72	0.77	−0.77	0.00

Negative values indicate carbon losses, positive ones - sources. *IC* inorganic carbon; *OC* organic carbon (Kuliński 2010)
[a] sum of organic carbon (0.45 Tg C year^{-1}) and black carbon (0.06 Tg C year^{-1}) loads

et al. (2003). The value of 13.56 Tg C year^{-1} obtained by the authors is based on the annual averages for water volumes and carbon concentrations. Thomas et al. (2003) assumed that the carbon concentrations are uniform for both water masses, and the values are close to data that characterize the North Sea. This assumption led to an overestimation of the export magnitude, since DIC concentrations (the main carbon species of the exchange) in the Baltic water are by 25–30% lower than those observed for the North Sea (Thomas and Schneider 1999; Prowe et al. 2009).

The third largest flux of carbon is carbon deposition to the sediments. It is the most essential mechanism that results in carbon resources depletion in the Baltic Sea (Table 5.4). The value, quantified to be −3.78 Tg C year^{-1}, is equal to 23% of all carbon losses from the basin, constituing at the same time the fundamental mechanism for the OC export from the Baltic, and representing 66% of the total organic carbon export. Carbon burial in the sediments is significantly reduced due to diffusion of dissolved carbon species from interstitial water to the overlaying water column (Fig. 5.1; Table 5.5). The return flux from the sediments is an effect of organic matter decomposition and mineralization processes that lead to 25–51% depletion of organic carbon accumulated in the bottom sediments, depending on the Baltic region. The lowest values of the return flux are observed for the Gotland Deep, the Gulf of Finland and the Gulf of Riga sediments, in contrast to the highest values for the Arkona Basin.

Organic matter fraction subjected to the decomposition and mineralization within the Baltic sediments is classified as labile and semi-stable. The diffusion flux has been estimated based on DIC and DOC concentration gradients within the uppermost 8 cm of the sediment cores collected in the key depositional areas of the sea. Assuming the average sedimentation rate for the deposition regions to be 2 mm year^{-1} (Pempkowiak 1991; Staniszewski 2005), the above conclusions concern the sediment layer that had been deposited within the last 40 years. However, the labile carbon fraction is subjected to decomposition taking place

Table 5.5 Summary of the annual averages for carbon accumulation in sediments in different regions of the Baltic, together with the magnitude of DIC and DOC return fluxes

Depositional area	Annual C accumulation in sediments (Tg year^{-1})	C return flux from sediments (Tg year^{-1})			C loss from sediments (%)
		DIC	DOC	DIC + DOC	
Arkona Basin	0.064	0.031 (94%)	0.002 (6%)	0.033	51
Bornholm Deep	0.764	0.237 (95%)	0.012 (5%)	0.249	32
Gdansk Deep	0.729	0.220 (95%)	0.012 (5%)	0.232	32
Gotland Deep	1.130	0.243 (91%)	0.025 (9%)	0.268	25
Gulf of Riga	0.036	0.008 (89%)	0.001 (11%)	0.009	25
Gulf of Finland	0.038	0.008 (89%)	0.001 (11%)	0.009	25
Gulf of Bothnia	1.018	0.297 (87%)	0.042 (13%)	0.339	33
Sum	3.779	1.044	0.095	1.139	30

Carbon loss is calculated as a ratio of the combined DIC and DOC return fluxes to carbon deposition (Kuliński and Pempkowiak 2011b)

over days/weeks time period from the moment of reaching the floor of the basin (Rullkötter 2006). This way, by applying DIC and DOC concentrations gradients, it is feasible to quantyfy long-term, time-averaged values for carbon losses from the sediments and the organic matter transformations in the course of an early sediments digenesis. The obtained results (Tables 5.4 and 5.5) point to sedimentation as an efficient mechanism of the organic carbon export from the water column. Simultaneously, the observed disparity of the carbon fraction returning to the water column, in the face of the total organic matter reaching the bottom (Table 5.5), suggests the diversity of conditions, mainly redox, in different depositional regions. Relevant, here, become aerobic/anoxic conditions prevailing in the near-bottom zone. These conditions depend on the depth of basin as well as the magnitude and frequency of well oxygenated, dense water inflows from the North Sea. The biggest contribution (51%) of the return flux is observed for the Arkona Basin—relatively shallow (50 m) and influenced by the North Sea water inflows due to the region's proximity to the Danish Straits (BACC Author Team 2008; HELCOM 2009). Due to the rare inflows from the North Sea, the Bornholm Deep and the Gdansk Deep reveal smaller (32%) contribution of the carbon return flux; additionally, these regions are characterized by periodic oxygen deficit conditions in water, while the sediments are anoxic permanently (Voipio 1981; BACC Author Team 2008; Lass and Matthäus 2008; HELCOM 2009). The lowest proportion of the returning carbon flux is found for the Gotland Deep.

Interestingly, along with decreasing organic matter decomposition indicator in the sediments, the contribution of DOC in the carbon return flux from sediments increases (Table 5.5). For the southern Baltic, its value ranges from 5 to 6% of the total diffusion flux, while it reaches 9% for the Gotland Deep. This correlation confirms a substantial influence of redox conditions on the organic matter decomposition dynamics in the sediments. In the case of the Gulf of Bothnia, despite of its well oxygenated environment (HELCOM 2009), the level of the organic matter decomposition is comparable with those observed for the Bornholm Deep and the Gdansk Deep sediments (Table 5.5). Additionally, DOC contribution in the return carbon flux is the largest (13%) for the Gulf of Bothnia, which indicates greater contribution of the stable, compared to semi-stable, organic matter fraction being deposited in this region. The reason for this situation is a low primary productivity occurring in the Gulf of Bothnia (Wasmund et al. 2001), with a simultaneous considerable organic matter load delivered with river runoff, characterized by a significant contribution of the biochemically stable organic matter (Rachold et al. 2004; Rullkötter 2006).

The designated range of carbon losses from the Baltic Sea sediments (25–51%) (Table 5.5) is comparable with estimations made for other seas. For the Chukchi Sea, the organic matter mineralization in sediments is estimated at 44–52% (Stein and Macdonald 2004), and for the Barents Sea—at 20–40% (Zaborska 2007). The literature data compiled by Rullkötter (2006) point at mineralization of the deposited organic matter at levels about 55% for the Aarhus Bay, Danmark (Jørgensen 1996), 24–34% for the Cape Lookout Bight (Martens et al. 1992) and 20–90%, globally, for shelf seas and estuaries (Henrichs and Reeburgh 1987). Noticeable discrepancies among the data are related, most likely, to different oxygen conditions prevailing in the investigated areas, and, unfortunately, different techniques used to obtained the results. This also applies to the approach taken when estimating diffusion fluxes, as well as analysing carbon loss of the organic matter collected in sediment traps (Rullkötter 2006). The contribution of the DOC flux in the total carbon return flux from the sediments for the Baltic Sea (Table 5.5) is also comparable with DOC assessed for other locations. Burdige and Homstead (1994) calculated the DOC flux contribution for the Chesapeake Bay, to range from 3 to13%. For the Californian coast, Burdige et al. (1999) estimated the value at 5.4–9.5% of the total carbon released from the sediments. Similar values (8 and 14%) were obtained by Holcombe et al. (2001) for the Pacific coast of Mexico and by Papadimitriou et al. (2002) for the western coast of the British Isles.

The organic carbon loss from the Gdansk Deep sediments, determined using the first Fick's law of diffusion, amounts to 19.1 g C m^{-2} $year^{-1}$ (Table 4.6) a value twice smaller than the potential returning flux of carbon estimated for the experiment involving sediment incubation (35.5 g C m^{-2} $year^{-1}$). This result may suggest that the influence of incubation temperature on the intensity of organic matter mineralization and decomposition processes in the sediments is higher than previously assumed. The incubation experiment clearly reveals the dynamics of organic carbon concentration changes within the sediments. The experiment allows to detect two stages of decomposition and mineralization of semi-stable organic matter. During the first one, lasting for about 15 years, a faster organic matter mineralization is observed. During the following stage, that lasts for additional 40–45 years, the rate of mineralization is slower. The total time required for sedimentary organic matter decomposition and mineralization is equal to 55–60 years. A similar time frame is suggested by Emeis et al. (2000), with pronounced increase in the organic carbon deposition rates observed since 1950, a feature attributed to an increased carbon sedimentation as a result of eutrophication of the southern Baltic. On the other hand, as highlighted by the authors, this increase may have its source in qualitative changes of the recently deposited organic matter involving the matter stability.

Our results indicate clearly a need to consider mineralization of organic matter that occurs in the course of sediments early digenesis, when quantifying the organic carbon burial in the Baltic sediments. The organic matter decomposition and mineralization are important factors reducing the organic matter load that is buried in sediments eventually.

5.3 The Budget Uncertainty

The Baltic Sea trophic state has been evaluated based on the carbon budget. The accuracy of the budget depends on uncertainties that the budget components are charged with. Analysis of the individual components reveal three basic fluxes that shape the Baltic carbon cycle: carbon exchange with the North Sea, carbon inflow with the rivers, carbon deposition to sediments and carbon return flux from sediments. The closing term is exchange of CO_2 between seawater and the atmosphere The uncertainties assigned to the fluxes are decisive for the net CO_2 exchange between the seawater and the atmosphere accuracy.

The net carbon export from the Baltic to the North Sea, estimated at −7.67 Tg C year^{-1}, is a sum of two fluxes characterized by opposite directions: negative—the Baltic carbon export to the North Sea, and positive—the carbon import to the Baltic from North Sea. These fluxes are calculated as a product of the flowing water volume and the adequate carbon species concentration. Thus, the highest year-to-year uncertainty that the net result is charged with, is related to the variabilities of the volume component and the concentrations components. The year-to-year variability of the net exchange between the Baltic and the North Sea amounts to 32.3% (BACC Author Team 2008; Omstedt et al. 2004). As the estimate of the concentrations uncertainties, the variabilities in relation to time and space have been considered. For the North Sea and the Baltic Sea concentrations of DIC, the representativeness is based on the accuracy of the measurements, established experimentally as 1.6 and 1.5%, respectively. Consecutively, as the representativeness of DOC concentrations, the maximum error of 11.5% for a single analyses has been adopted. Based on this data, the uncertainty for the net carbon exchange between the Baltic and the North Sea has been computed according to Eq. 5.4. The uncertainty estimated this way is no larger than 2.49 Tg C year^{-1}, which corresponds to the relative uncertainty not exceeding 32.5% of the carbon flux.

$$\sigma_{EN} \leq \sqrt{(\sigma_V \cdot EN)^2 + (\sigma_{DIC_B} \cdot E_{DIC_B})^2 + (\sigma_{DOC_B} \cdot E_{DOC_B})^2 + (\sigma_{DIC_P} \cdot I_{DIC_P})^2 + (\sigma_{DOC_P} \cdot I_{DOC_P})^2} \quad (5.4)$$

where:

σ_{EN} absolute error of the carbon net export from the Baltic to the North Sea;

σ_V year-to-year variability of the water net exchange, equal to 32.3% (BACC Author Team 2008 after Omstedt et al. 2004);

EN absolute value of the net carbon exchange [Tg C year^{-1}] (Table 5.4);

σDIC_B representativeness of the Baltic DIC concentration, equal to 1.5% (Thomas and Schneider 1999);

$EDIC_B$ absolute value of the Baltic DIC export [Tg C year^{-1}] (Table 5.4);

σDOC_B representativeness of the Baltic DOC concentration, equal to 11.5%;

$EDOC_B$ absolute value of the Baltic DOC export [Tg C year^{-1}] (Table 5.4);

σDIC_P representativeness of the North Sea DIC concentration, equal to 1,6% (Prowe et al. 2009);

$IDIC_P$ absolute value of the North Sea DIC import [Tg C year^{-1}] (Table 5.4);

σDOC_P representativeness of the North Sea DOC concentration, equal to 11.5%;

$IDOC_P$ absolute value of the North Sea DOC [Tg C year^{-1}] (Table 5.4);

The total carbon load transported to the Baltic Sea with rivers (10.90 Tg C year^{-1}, Table 5.4) is calculated as a product of the water volume and both the organic and inorganic carbon concentrations. Therefore, the uncertainty of estimates is calculated as a sum of year-to-year variability of both. The variability of the river water volume flowing into the Baltic is estimated at 28.6% (BACC Author Team 2008 after Omstedt et al. 2004). As representative for both organic and inorganic carbon concentrations, the maximum error of 11.5% obtained for a single analysis has been assumed as a result of DOC measurement series. Thus, the uncertainty of the estimates for the total carbon load carried to the Baltic with rivers is not greater than 3.36 Tg C year^{-1}, which corresponds to uncertainty not exceeding 30.8% of the total load, according to the equation:

$$\sigma_R \leq \sqrt{(\sigma_V \cdot R)^2 + (\sigma_{st} \cdot R)^2} \tag{5.5}$$

where:

σ_R error of the total carbon load carried by rivers

σ_V year-to-year variability of the river water inflows to the Baltic equal to 28.6% (BACC Author Team 2008 after Omstedt et al. 2004)

R total carbon load carried by rivers to the Baltic [Tg C year^{-1}] (Table 5.4)

σ_{st} representativeness of TIC and TOC concentrations in rivers, equal to 11.5%

In order to determine the uncertainty limits for the organic carbon deposition in the sediments and the return flux of the dissolved carbon species from the sediments, an error of 28% established as a measure for accuracy based on a single sediments core has been accepted. The value determined by Emeis et al. (2000) for the Gotland Deep is the largest among all depositional areas considered and investigated for the Baltic Sea. Based on the above, the uncertainty estimation for

Table 5.6 Summary of the maximum uncertainty values of the Baltic Sea individual carbon budget components

Flux	Carbon load (Tg C year^{-1})	Uncertainty (Tg C year^{-1})	Uncertainty (%)
Rivers	10.90	3.36	30.8
North Sea	−7.67	2.49	32.5
Sediments	−2.64	1.10	41.7
Precipitation	0.57	0.07	13.3
Wastewater	0.04	i. u.	<7
Fisheries	−0.06	i. u.	<8

i. u.—insignificant uncertainty (< 0.01 Tg C year^{-1})

the carbon deposited to the sediments (−3.78 Tg C year^{-1}; Table 5.4) amounts to ± 1.06 Tg C year^{-1}, while for the return flux of carbon from sediments it is ± 0.32 Tg C year^{-1} (1.14 Tg C year^{-1}). Therefore, the net carbon export to the sediments (−2.64 Tg C year^{-1}) is charged with uncertainty equal to 1.10 Tg C year^{-1} (Tables 5.4 and 5.6).

The year-to-year variability that characterizes the precipitation reaching the Baltic Sea is estimated at 13.3% (BACC Author Team 2008 after Omstedt et al. 2004). This value is adopted as the error that the carbon load accompanying precipitation (0.57 Tg C year^{-1}, Table 5.4) is charged with. Thus the estimate of uncertainty does not exceed 0.07 Tg C year^{-1} (Table 5.6). Based on the source data (HELCOM 2004; ICES 2008) for the carbon loads originating from the point sources (0.04 Tg C year^{-1}, Table 5.4) and fisheries (−0.06 Tg C year^{-1}, Table 5.4), the year-to-year variability has been estimated to be lower than 7% and 8%, respectively. The absolute errors of the last two budget components are thus identified as not significant for the budget stability.

Maximum uncertainty values that individual components of the budget are charged with are summarized in Table 5.6. Their contributions, in the investigated carbon fluxes, are comparable to those observed for analogical estimates for other locations, derived both experimentally or theoretically (e.g., Chen et al. 2003; Karl et al. 2003; Stein and Macdonald 2004). Additionally, a characteristic feature of this type of estimates is that the uncertainty estimates for the obtained results are seldom applied, especially with regards to the year-to-year variability (e.g., Chen et al. 2003; Macdonald et al. 2004; Naidu et al. 2004; Stein and Macdonald 2004; Emerson and Hedges 2008).

Since the main carbon fluxes in the Baltic Sea constitute sinks or sources of carbon, the absolute value of their sum, that conditions the direction and magnitude of the net CO_2 exchange between seawater and the atmosphere, is considered insignificant when compared to the absolute values of the individual components of the sum. Therefore, the theoretical value of the uncertainty assigned to the net CO_2 exchange, amounting to the square root of the sum of the squared errors of all components, is significantly larger (4.32 Tg C year^{-1}) than the absolute value of the actual flux through the seawater/atmosphere interface (1.14 Tg C year^{-1}).

Nonetheless, it may be safely assumed that in the course of time, the Baltic Sea trophic status can switch from heterotrophic to autotrophic and vice versa. Considerations should also involve feedbacks that bind the individual carbon fluxes—especially those dependant on the water flows. An increase of river water volume, discharged to the Baltic Sea, directly determines the basin water level (Cyberski and Wróblewski 2000) that is the driving force of the Baltic water transport to the North Sea and for blocking inflows from the North Sea in the same time (BACC Author Team 2008; Lass and Matthäus 2008). The reciprocity of carbon fluxes originating from the water flow has been well illustrated based on the Baltic-North Sea exchange during the subsequent years (Sect. 4.1). When compared to the successive annual cycles, years 2002/2003 are characterized by higher water export and import values (Table 4.1) as well as DIC (Table 4.2) and DOC fluxes (Table 4.3). Such a state is associated with a major water inflow from the North Sea that took place in 2003 (BACC Author Team 2008), which influenced the annual water import and dissolved DIC and DOC fluxes into the Baltic. On the other hand, the events like that increase the export to the North Sea of both dissolved DIC and DOC, due to increased water outflow necessary in order to achieve a hydrodynamic balance between the two basins. To determine the limit for the net CO_2 exchange between the Baltic and the atmosphere, it has been assumed that all estimate uncertainties of the year-to-year variabilities of the carbon fluxes dependent on water volumes compensate one another. Based on this assumption, the net CO_2 exchange between the seawater and the atmosphere has been estimated at -1.14 ± 1.69 Tg C year^{-1}. The obtained error provides the estimates that rely on the carbon fluxes mass balance (Anderson et al. 1998; Chen et al. 2003; Borges 2005). Despite their significant uncertainty, the presented estimates of CO_2 absorption/emission to the atmosphere classify the Baltic Sea as a basin characterized by, a close to, neutral balance of the annual CO_2 exchange, with a little skewness towards the export (emission).

5.4 Deposition and Origin of Organic Matter to the Baltic Sediments. The Case of Cores Dated with ^{210}Pb and ^{137}Cs

Introduction. Considerations presented in this chapter are based on the comprehensive study performed by Szczepańska et al. (2011).

The Baltic Sea has been described as an autotrophic semi-enclosed brackish sea (Thomas et al. 2003, this volume). Considerable amounts of nutrients, mostly from agriculture and industry, enter the sea with river run-off, making the Baltic Sea one of the most productive marine ecosystems (Emelyanov 1995; Gudelis and Jemielianow 1982; HELCOM 2006; Thomas et al. 2003). The most important sources and sinks of the organic matter in the Baltic Sea are: primary production, river runoff, exchange with the North Sea, and deposition to sediments (Wasmund et al. 2001). However, it must be kept in mind that the Baltic Sea is a net source of organic matter for the North Sea (Kuliński et al. 2011).

A large proportion of the Baltic Sea catchment area incorporates agricultural land that is a source of organic matter and nutrients loads (Emelyanov 1995; Gudelis and Jemielianow 1982). The increase of nutrients concentrations intensifies the "biological pump". The "biological pump" mechanism causes CO_2 absorption by phytoplankton and its assimilation to organic matter. A large part of organic matter, both brought to the sea with river runoff (allochtonous organic matter) and produced within the sea (autochtonous organic matter), is incorporated into the sediments (Kuliński and Pempkowiak 2008; Zaborska et al. 2008). Existing data are insufficient to quantify accurately the load of organic carbon that is accumulated in sediments. Preliminary estimates indicate that the amount of organic carbon deposited to the sea bottom equals about 3.78 Tg C_{org} yr^{-1} (Kulinski and Pempkowiak 2011a, b). However, the exact value is unknown (Sect. 5.3). Estimates are uncertain due to limited amount of data on both sediment accumulation rates and organic carbon concentrations in bottom sediments. However, it is obvious that the deposition of carbon species to sediments is an important sink of carbon in the Baltic sea. The obvious reasons have been already mentioned. These are the high primary productivity, relative shallowness of the sea, large loads of dissolved and particulate organic matter discharged to the sea with the river run-off, and rather low average temperature.

The brackish nature of the Baltic is due to the combination of the high river water discharge and the limited inflows of saline water from the North Sea. Inflows of saline water lead to a stable stratification in the basin water column. The water budget of the sea is further influenced by precipitation that exceeds evaporation. Halocline, located at a depth of 60–80 m, is a result of limited mixing of water masses. This, combined with the progressive eutrophication, are the causes of oxygen deficits. Thus, the Baltic water below the halocline is depleted of oxygen, frequently even comprising hydrogen sulfide. This leads to limited mineralization of organic matter in the time interval required for deposition of organic matter to the sea bottom. The hydrolysis and mineralization of organic matter continues in the sediments in the course of early diagenesis. The time necessary for the full stabilization of organic matter has been estimated to be 50–60 years (Sect. 4.2).

As a consequence, load of land-derived organic matter, that is largely composed of species biochemically resistant to degradation (Pempkowiak and Kupryszewski 1980), will be readily deposited to the sea bottom. This flux of organic matter will be accompanied by fresh autochtonous organic matter deposited to the sea bottom because there is no time for it to be mineralized in the shallow column of cool water. On the one hand, the latter flux will contribute to the deposition of organic carbon to the sea bottom. However, the fresh, labile organic matter will be a subject of subsequent biochemical degradation in the course of early diagenesis. It could be speculated that relatively fast moving mineralization/hydrolysis mentioned in the previous chapter concerns the fresh portion of the deposited organic matter, while the longer, several decades long, mineralization concerns the more biochemically resistant, land-derived fraction of the organic matter deposited to the bottom sediments (see Sect. 4.3 for discussion regarding decomposition and mineralization of organic matter in the sediments).

The two above-mentioned phenomena should be causing alternation of the percentage contribution of autochtonous and allochtonous organic matter in the sediments. The uppermost sediment layers should be enriched with the autochthonous organic matter fraction, while the subsurface sediments should have a larger proportion of allochtonous fraction. More so, as the primary productivity in the Baltic sea nowadays is more than twice that characterizing the sea some seventy years ago. The reason is a strong eutrophication of the sea and large primary production resulting from the seawater enrichment with nutrients (Łysiak-Pastuszak 2000). The latter phenomenon, of course, enforces the former one. The described mechanism is, in fact, not significant for the carbon budget in the Baltic, as no differentiation between the autochtonous and allochtonous organic matter fluxes within the sea needs not to be considered. Nevertheless, if proven, the quantification of the sedimentary organic matter origin might contribute to the understanding of carbon cycling within the sea, and support the concepts the carbon budget in the Baltic is based on.

In this section, results of work aimed at recognition of organic matter deposition and provenience in sediments are reported. Central basins of the sea were selected as the study area since most sedimentation takes place in the deep depositional basins of the Baltic Sea— the Gotland Deep, and the Gdansk Deep (Szczepańska et al. 2011).

Considerations regarding carbon are accompanied by nitrogen, an important biogenic element that is believed to be a limiting factor for the primary production of organic matter in the Baltic. Although the sources of nitrogen to the Baltic are well established, deposition to sediments still requires further quantification (Łysiak-Pastuszak 2000; Wulff et al. 2001; Thomas et al. 2003). Using mass sediment accumulation rates, derived from ^{210}Pb and ^{137}Cs profiles, and concentrations of organic carbon and total nitrogen, accumulation rates of both these elements in the sediments of the Baltic Sea were calculated. The study was carried out on sediment cores collected from depositional areas of the sea. The results are deemed essential for carbon and nitrogen budget calculations in the Baltic Sea.

The details of methodology used for this study are described in Appendix (Sect. A.3)

Sediment accumulation rates. The highest total ^{210}Pb ($^{210}Pb_{tot}$) activity concentrations were measured at station VI, ranging from 906 Bq/kg at the sediment surface to 211 Bq/kg at lower sediment section. The lowest total ^{210}Pb activity concentrations are measured at station I. They range from 172 Bq/kg to 26 Bq/kg (Fig. 5.2a). The supported $^{210}Pb_{supp}$ activity concentration was calculated as the mean of activities in layers below the decay zone, with constant activity. $^{210}Pb_{ex}$ activity concentration was calculated as the difference between total $^{210}Pb_{tot}$ and $^{210}Pb_{supp}$ activities:

$$^{210}Pb_{ex} = {}^{210}Pb_{tot} - {}^{210}Pb_{supp} \tag{5.6}$$

Sediment accumulation rates vary in the range from 0.06 cm yr^{-1} at station IV to 0.16 cm yr^{-1} at station VI. Mass sediment accumulation rates, below the depth

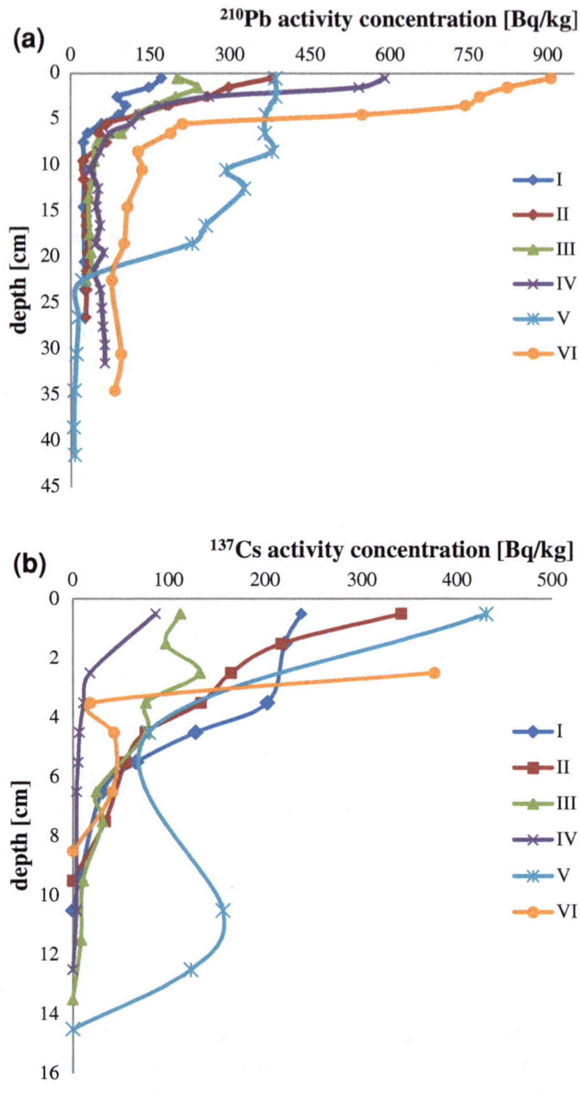

Fig. 5.2 **a** The ^{210}Pb activity concentrations against sediment depth in the cores, **b** The ^{137}Cs activity concentrations against sediment depth in the cores (Szczepańska et al. 2011)

of surface sediment mixing, vary from 159 to 534 g m^{-2} yr^{-1}. The highest mass sediment accumulation rate was found at station I (534 g m^{-2} yr^{-1}) and station III (461 g m^{-2} yr^{-1}), while the mass sediment accumulation rate at station II reached a value of just 160 g m^{-2} yr^{-1} (Table 5.7).

The results are well in the ranges of sedimentation rates reported earlier for the Baltic (Hille et al. 2006; Pempkowiak 1991; Szczepanska et al. 2009; Suplinska 2008; Widrowski and Pempkowiak 1986).

The activity concentrations of ^{137}Cs in sediments differ depending on sampling site and sampling depth (Fig. 5.2b). The highest concentration is observed at station V, localized in the Gdansk Deep, amounting to about 430 Bq/kg in the

Table 5.7 Sediment accumulation rates, mass sediment accumulation rates, organic carbon accumulation rates and nitrogen accumulation rates in the investigated sediments (Szczepańska et al. 2011)

Sediment core	Sediment accumulation rate ($cm\ yr^{-1}$)	Mass sediment accumulation rate ($g\ m^{-2}\ yr^{-1}$)	Organic carbon accumulation rate ($g\ C\ m^{-2}\ yr^{-1}$)	Accumulation rate of nitrogen ($g\ N\ m^{-2}\ yr^{-1}$)	Ratio of C and N accumulation rates
I	0.14	534	34	4.4	7.7
II	0.11	198	23	2.3	10.0
III	0.10	461	43	5.0	8.6
IV	0.06	320	18	2.3	7.8
V	0.59	410	27	3.4	7.9
VI	0.16	874	75	8.8	8.5

surface layer, while the lowest—at station IV, where the activity concentration is about 80 Bq/kg at surface sediments. Ranges of radiocesium activity concentration in the surface sediments reported earlier (Pempkowiak 1991; Suplinska 2008) agree well with our results.

Accumulation rates of organic carbon. Organic carbon concentrations measured in all the six investigated sediment cores are presented in Fig. 5.3a. The highest organic carbon concentration (C_{org}), ranging from 10.13% at the surface to 4.56% in deeper layers, is observed at station II, situated on the southern side of the Gotland Deep. At station IV, the C_{org} concentration reached value as low as 5.47% at the surface layer and decreased to just 3.73% in deeper layers.

Organic carbon accumulation rates calculated as a product of the mass sediment accumulation rates and the organic carbon concentration vary in the range from 18 $g\ C\ m^{-2}\ yr^{-1}$ at station IV to 75 $g\ C\ m^{-2}\ yr^{-1}$ at station VI (Table 5.7).

The average organic carbon accumulation rate observed at stations II and III (Gotland Deep) is about 33 $g\ C\ m^{-2}\ yr^{-1}$. The value is within ranges obtained by Emeis et al. (2000) (about 27 $g\ m^{-2}\ yr^{-1}$) and Struck et al. (2000) (about 30 $g\ m^{-2}\ yr^{-1}$), and reported in this study, for the southern Baltic.

The organic carbon accumulation rate calculated for station V amounts to 27 $g\ C\ m^{-2}\ yr^{-1}$. The value is smaller than the literature values for the Gdańsk Deep (Emeis et al. 2000, about 60 $g\ m^{-2}\ yr^{-1}$), although the mass sediment accumulation rate value is similar to the literature values for this region (MAR at station V – 410 $g\ m^{-2}\ yr^{-1}$, MAR obtained by Pempkowiak 1991, about 510 $g\ m^{-2}\ yr^{-1}$). The differences are caused by low levels of organic carbon concentration, measured in sediment samples from the core number V, amounting to about 6.48% in the surface and just 2.87% in deeper segments of the sediment core, caused by location of the station at the margin of the Gdansk Deep.

Profiles of organic carbon content in recent bottom sediments of the Baltic Sea indicate that organic matter concentration decreases with sediments depth. This is attributed to the increased organic matter deposition caused by eutrophication (Emeis et al. 2000; Voss et al. 2000). Another important factor contributing to this phenomenon is mineralization of labile part of organic matter (see Sect. 4.3 for

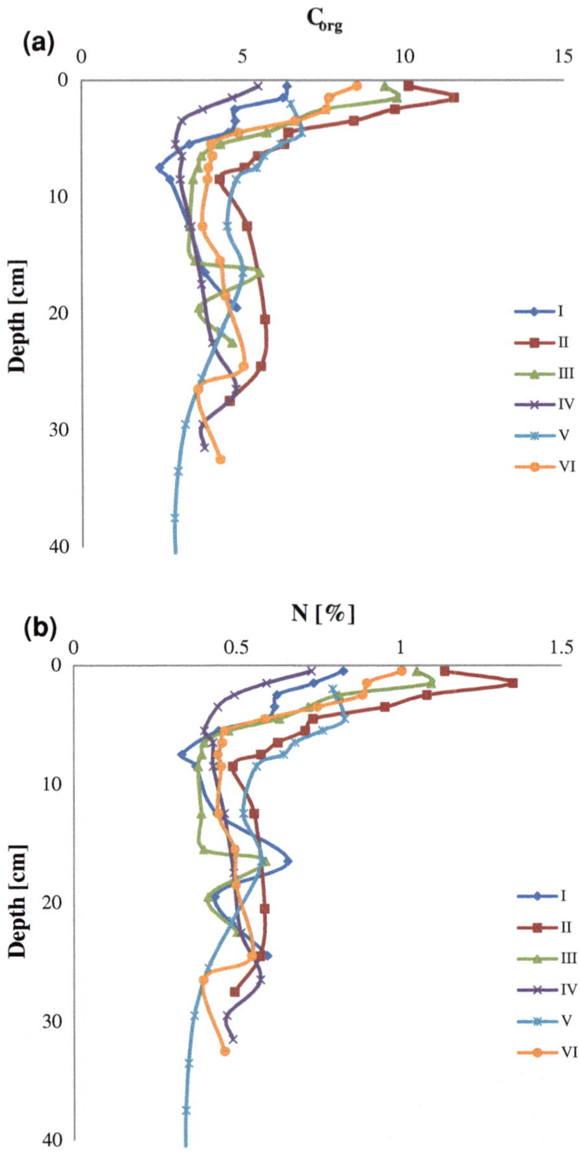

Fig. 5.3 **a** The organic carbon concentration against sediment depth in the cores, **b**. Total nitrogen concentration against sediment depth in the cores (Szczepańska et al. 2011)

discussion of the phenomenon). Such a trend is observed generally at stations II, III, IV and V. At station III the typical decrease is followed by an increase in the subsurface sediment layers. This increase may suggest that 100–150 years ago, at this particular location, there was much higher input of organic matter that has been buried permanently in sediments. Another explanation might be a different origin and, related to this, different properties of organic matter deposited there due to resuspension and redeposition processes.

Organic carbon concentration profiles and total nitrogen concentration profiles at all stations except station I exhibit similar shapes. This means that nitrogen, carbon and organic matter concentration changes in the corresponding sediment layers are governed by similar processes. This is clearly seen at sediment core from station III, where increases of nitrogen and organic carbon concentrations are measured at exactly the same sediment layer (16–17 cm), indicating untypical origin of sediments constituting the layer.

Accumulation rates of nitrogen. Total nitrogen concentrations in the investigated sediment samples range from 1.34 to 0.34%. The highest N_{tot} concentration was measured at station II, amounting to 1.34% in the surface layer and 0.48% in deeper part of the core. The lowest total nitrogen concentration was measured in sediment core collected at station IV (0.73–0.42%) (Fig. 5.3b).

Accumulation rates of nitrogen expressed in g N m^{-2} yr^{-1} were calculated as a product of the mass sediment accumulation rates and the nitrogen concentrations measured in sediments. The lowest nitrogen accumulation rate is observed at stations II and IV (2.3 g N m^{-2} yr^{-1}). The highest nitrogen accumulation rate is observed at station VI (8.8 g N m^{-2} yr^{-1}) (Table 5.7).

As one can see, the same trends are observed for accumulation of total nitrogen and organic carbon. The lowest and the highest values of organic carbon accumulation rates have also occurred in sediments from stations II and VI. However, ratios of C/N accumulations differ considerably, indicating perhaps that not all labile organic matter is actually mineralized in the surface sediments. The matter of C/N ratios is discussed, in some detail, below. If deposition of carbon to the Baltic sediments equal to 3.8 TgC/annum (Table 5.5), and the ratio of carbon flux to nitrogen flux to sediments equal to 9 (Table 5.7), are adopted, nitrogen deposition to sediments ammounts to some 0.4 Tg/annum. Thus the Baltic sediments act as substantial sink of nitrogen.

Origin of organic matter in the Baltic sediments. There are several indicators that may be used to assess sedimentary organic matter provenance. The most valuable ones are based on molecular composition of organic substances. Moreover, the ratio of C/N abundance in organic matter is frequently used for the purpose of differentiating land (C/N $>$ 12) versus marine (C/N $<$ 10) origin. Redfield ratio of marine phytoplankton biomass (C/N $=$ 6) is much smaller than that of sedimentary organic matter originating from pure marine sources. This indicates that preferential mineralization of organic nitrogen takes place in the course of sedimentation and early diagenesis. The C/N ratios encountered in the investigated cores vary in the range from 7.4 to 9.6 (Table 5.8). This would indicate the predominantly marine character of sedimentary organic matter in the investigated sediments. However, due to the unknown molecular composition of both marine and land components, just qualitative assessment as to the organic matter origin can be performed. Moreover, the ratios measured in the sediments may be biased due to the differing mobility of organic matter mineralization products (CO_2 and $NH_3{*}H_2O$). Carbon dioxide largely diffuses out of the sediments to the overlying water, while ammonia is partly sorbed on sediments particles and thus retained in sediments (Emeis et al. 2000). Thus the C/N ratio

Table 5.8 The C_{org}/N_{tot} ratios, $\delta^{13}C$ and terrestrial organic matter proportion in the surface (<10 cm) and subsurface (>20 cm) segments of the investigated cores (Szczepańska et al. 2011)

Sediment core	Layers	C/N ratio $\pm \sigma^{a}$	$\delta^{13}C \pm \sigma^{a}$ [‰]	Terrestrial organic matter contribution [%]
I	Surface	8.6 ± 0.2	−24.70 ± 0.08	45.0
	Subsurface	9.6 ± 0.5	−24.68 ± 0.98	44.6
II	Surface	8.8 ± 0.1	−24.56 ± 0.20	42.7
	Subsurface	9.5 ± 0.2	−25.63 ± 0.10	60.5
III	Surface	9.1 ± 0.2	−24.70 ± 0.18	45.0
	Subsurface	9.0 ± 0.3	−26.36 ± 0.36	72.7
IV	Surface	7.4 ± 0.3	−24.78 ± 0.12	46.5
	Subsurface	8.0 ± 0.3	−24.63 ± 0.27	44.2
V	Surface	8.3 ± 0.1	−24.61 ± 0.22	43.5
	Subsurface	8.8 ± 0.2	−25.81 ± 0.08	63.5
VI	Surface	8.6 ± 0.2	−24.45 ± 0.21	40.8
	Subsurface	9.0 ± 0.2	−25.53 ± 0.06	58.8

[a] standard deviation

can be used for quantitative assessment of organic matter origin (Table 5.8), but cautiously.

It is obvious, however, that C/N ratios in the uppermost segments of the cores (<10 cm) are much lower than those in the subsurface segments (>20 cm). The border between higher (subsurface) and smaller (surface) ratios is placed at the depths corresponding to deposition period 1930–1950, close to the time when eutrophication in the Baltic has started (Łysiak-Pastuszak 2000). For five of the six investigated cores these differences are statistically significant. The average C/N ratio in the surface sediments is 8.4. There is one core (IV) with C/N ratio (7.4) significantly lower than average, and one core (III) with C/N ratio (9.1)—significantly higher. In the latter case, the differences seem to indicate increased contribution of terrestrial organic matter, originating possibly from the Bay of Riga while the former—the accumulation of large quantities of sedimentary material rich with autochthonous organic matter, possibly due to redeposition and/or sediment slides on steep sea bottom.

Stable carbon isotopes were also measured in the samples originating from the sediment cores and are presented as $\delta^{13}C$ in Table 5.8. The $\delta^{13}C$ values measured in the sediment samples ($\delta^{13}C_s$) range from −24.3 to −26.6‰. This is in good agreement with the earlier published data (Voss et al. 2005). The measured values indicate that the organic matter in sediments is a mixture of terrestrial ($\delta^{13}C_t$ equal to −28‰) and marine ($\delta^{13}C_m$ equal to −22‰, Boutton 1991) materials. The $\delta^{13}C$ values are frequently used for quantitative assessment of terrestrial organic matter (T%) contribution in the coastal marine environment (Voss et al. 2005). An end member approach is applied for this purpose, based on the following formula:

$$T(\%) = \frac{\delta^{13}C_S - \delta^{13}C_m}{\delta^{13}C_t - \delta^{13}C_m} \times 100(\%) \tag{5.7}$$

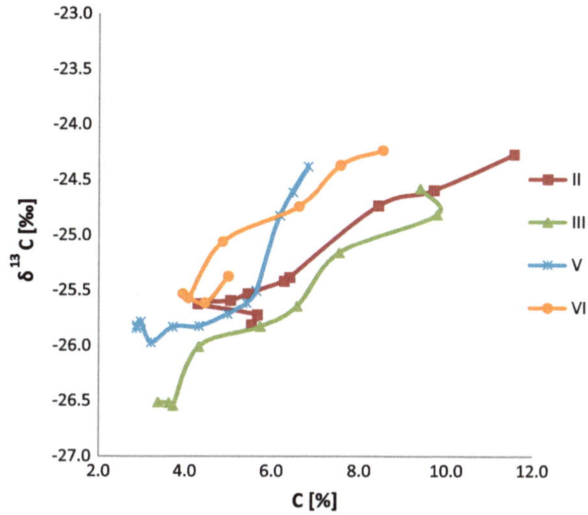

Fig. 5.4 The relationship between $\delta^{13}C$[‰] and C[%] concentrations in the cores (Szczepańska et al. 2011)

Results of the so calculated terrestrial contribution to sedimentary organic matter are presented in Table 5.8. The obtained values, for the surface sediments, are surprisingly uniform, indicating that terrestrial organic matter constitutes between 40.8 and 46.5% of total organic matter in the surface sediments, independently of the location. Proportion of terrestrial organic matter in subsurface sediments ranges between 44.2 and 72.7%. Thus, it is both much larger and much more variable. This can be attributed to the recent eutrophication of the sea and increased production of autochthonous organic matter (Łysiak-Pastuszak 2000; Voss et al. 2000). This explanation is supported by the high correlation coefficient of the relationship between $\delta^{13}C$ and C concentration (Fig. 5.4). A general agreement between $\delta^{13}C$ and C/N ratios, namely decreasing $\delta^{13}C$ paralleling increasing values of C/N is also evident, and most likely caused by mineralization of nitrogen–rich organic matter.

Conclusions. The organic carbon deposited in marine sediments is an important part of the global carbon cycle. The knowledge concerning the role of shelf seas (including the Baltic Sea) in the carbon cycle has increased substantially, however the organic carbon accumulation rates in the Baltic sediments still require clarification. Organic carbon accumulation rates in six sediment cores collected in the sediment accumulation areas in the Baltic Proper were used for assessing the importance of sedimentary organic matter in carbon cycling in the Baltic sea. Mass sediment accumulation rates were based on ^{210}Pb method validated by ^{137}Cs measurements. The obtained results point to the Baltic Sea sediments as an important sink of organic carbon; the organic carbon accumulation rates range from 18 to 75 g C m^{-2} yr^{-1}. This is in good agreement with data used to assess carbon deposition to the Baltic sediments (Sect. 4.3) Large proportion (41–73%) of the deposited organic matter originates on land as derived from stable carbon isotopes ($\delta^{13}C$) composition. The proportion of the land derived organic matter gradient in the surface and subsurface sediments supports clearly the concept of carbon deposition to, and mineralization in, the Baltic sediments

5.5 Influence of the Climate Change on the Carbon Resources and Fluxes in the Batlic Sea

As described in Sects. 5.1 and 5.2, the carbon budget in the Baltic Sea and the impact of its individual components can alter substantially during the coming decades. There is a number of mathematical models simulating changes that can take effect before 2100 (e.g., Graham 2004; Meier 2006; Graham et al. 2007; BACC Author Team 2008; Dzierzbicka-Glowacka et al. 2010). Even though considered for the local scale (Baltic and its drainage area), these changes are not easily and precisely predictable since the determining factors are often of a global magnitude. They are related, to a large extent, to the global climate change caused mostly by the elevated emission of greenhouse gases. Among the gases, the major role is played by the anthropogenic CO_2 originating mostly from the fossil fuels combustion, cement production, and changes occurring within the biosphere (IPCC 2007); therefore, the mechanisms determining the anthropogenic CO_2 emissions are due to both, the socio-economic background, as well as the political developments. Due to the complexity of processes governing the anthropogenic CO_2 cycle in the environment, and the greenhouse gases concentrations in the atmosphere, it becomes difficult to predict the evolution of climate in both the global and local scales (IPCC 2001 and 2007). Nevertheless, the majority of predictions point at an increase in the global temperature, and the problems resulting from it (IPCC 2001 and 2007; BACC Author Team 2008).

One of the most frequently mentioned effects of the global climate change is a transformation in the Baltic Sea hydrological regime (Graham et al. 2007; BACC Author Team 2008). It is estimated that until 2100, the volume of water discharged by rivers will increase by as much as 15% (Graham 2004; BACC Author Team 2008). Still bigger increase, by 10–40%, might be observed for selected subbasins of the Baltic, e.g., the Gulf of Bothnia and the Gulf of Finland. In the northern Baltic, a change in river runoff seasonality is expected to occur, with a shift of its maximum from late spring to the winter/spring period. Causes of these changes are sought in milder winters, and thus in an ice and snow cover shrinkage within the boreal part of the Baltic, as well as elevated levels of precipitation (Graham et al. 2007). Investigations conducted for three Swedish rivers: Kalix, Torne and Råne have shown that the volume of water discharged to the Baltic is responsible, in up to 97%, for the magnitude of the riverine carbon load carried from the land (BACC Author Team 2008). Therefore, it is expected that with the 15% increase in river water runoff to the Baltic Sea, the carbon load carried with rivers will simultaneously increase by, at least, 15%. Currently, the carbon load is estimated at 10.90 Tg C year^{-1} (Table 5.4). Thus the average carbon flux reaching the Baltic with river runoff may increase to 12.53 Tg C year^{-1}. This estimate includes the IC and OC load discharged from both the Gulf of Finland and the Gulf of Bothnia drainage areas. OC is supposed to originate from an increased erosion of soil in the drainage area as an effect of elevated winter temperature. The temperature rise will cause the ground melting, allowing water and dissolved CO_2 to percolate within the porous medium, and causing prolonged weathering and organic matter extraction. Significantly smaller river runoffs, according to some studies, will be observed for the southern Baltic (Graham 2004; BACC Author Team 2008). The precipitation volume reduction is thought to be the main cause there. Along with smaller precipitation volume in the area, the number of extreme precipitation incidents is expected to increase. Therefore, a decrease of carbon discharged to the southern Baltic by rivers is expected due to diminished volume of the water, parallel to its increase driven by a strong erosion caused by events of intensive rainfalls and flooding. While it is rather difficult to estimate if these two counteracting factors will compensate one another, it is projected that the amplitude of the seasonal carbon loads carried with the continental rivers will increase.

The average increase of the river runoff to the Baltic will automatically lead to an increase of the water export to the North Sea and to a limited import of the North Sea water (Cyberski and Wróblewski 2000; BACC Author Team 2008), which in turn will result in an elevated carbon outflow to the North Sea. It should be taken into account that the river runoff as compared to the sea water is characterized by higher carbon concentrations, organic carbon in particular, thus even the elevated carbon export through the Danish Straits may not compensate the increase of carbon inflows with rivers.

An additional factor that will likely affect the carbon cycle during the coming decades is the coastline erosion occurring as an effect of rising water levels. Prognosis of 1 m water level increase for the south-eastern Baltic by the end of the century is deemed certain. This will lead to flooding of vast coastal areas, and incorporating into the Baltic carbon budget, loads of carbon deposited in soil (Pempkowiak et al. 2009).

The available estimates are contradictory when it comes to assessing the consequences of limited water inflows from the North Sea and the increase of river runoff (Gerlach 1994; Meier 2006; BACC Author Team 2008). On the one hand, mathematical modelling (Meier 2006; BACC Author Team 2008) suggests a probability of the salinity decrease, across the water column, leading to a stronger vertical stratification and halocline occurrence at unchanged depths. On the other hand Gerlach (1994) and BACC Author Team (2008) suggest that the halocline depth will lower as a result of river runoff increase, as well as due to less intensive events of the North Sea water inflows. The latter reason might increase organic matter mineralization in anoxic zones, particularly in the sediment deposition areas. Well oxygenated water would, in turn, provide conditions for intensified organic matter mineralization processes, that lead to releasing an additional carbon loads to the water column.

It is projected that the climate warming effects for the Baltic Sea area will exceed the average global warming before 2100 (BACC Author Team 2008). It has been estimated that the average temperature will increase by 3–5°C, and by 4°C for summer and winter seasons, respectively. Increased temperature of the surface water layer, combined with less saline Baltic surface waters, due to increased river runoff, can lead to significant changes in phytoplankton population structure, resulting, for instance, in limited spring bloom of coldwater diatom species, favoring at the same time cyanobacteria and dinoflagellata growth (Wasmund and Uhlig 2003; BACC Author Team 2008; Wasmund and Siegel 2008). Anticipated is also an increase in molar ratios of N/Si and N/P in seawater, as a result of increasing nutrient supplies delivered from the land and diminishing silica and phosphorus availability caused by their precipitation in the sediments driven by improved oxygen conditions (Gerlach 1994; Jansson 2001; BACC Author Team 2008). Under these circumstances, diatoms and cyanobacteria growth would be significantly altered. The projected decrease of pH induced by the atmospheric CO_2 could also lead to modification of species composition and succession of phytoplankton (BACC Author Team 2008). It is still unknown what changes of the Baltic Sea carbon cycle could be a consequence of the phytoplankton species composition shifts.

In conclusion, an elevated carbon inflow to the Baltic Sea, particularly composed of organic species, and an increase of carbon re-emission from the sediments, could raise levels of CO_2 in seawater causing increased emission to the atmosphere. The differences observed for the individual Baltic regions could also deepen. The Gulf of Bothnia, characterized by, already intensive, CO_2 degassing to the atmosphere, can become an even stronger emitter, while the southern part of the Baltic will remain the absorptive region. Less intensive changes within the southern Baltic will be determined by the primary productivity sustained at high levels, and the constant natural IC and OC export to the North Sea. Unknown is the extent of changes that may occur in the Baltic as a result of the trophic chain evolution.

References

Abril G, Etcheber H, Delille B, Frankignoulle M, Borges AV (2003) Carbonate dissolution in the turbid and eutrophic Loire estuary. Marine Ecol Prog Ser 259:129–138

Algesten G, Wikner J, Sobek S, Tranvik LJ, Jansson M (2004) Seasonal variation of CO_2 saturation in the Gulf of Bothnia: indications of marine net heterotrophy. Glob Biogeochem Cycles 18:GB4021

Algesten G, Brydsten L, Jonsson P, Kortelainen P, Löfgren S, Rahm L, Räike A, Sobek S, Tranvik L, Wikner J, Jansson M (2006) Organic carbon budget for the Gulf of Bothnia. J Marine Syst 63:155–161

Anderson LG, Olsson K, Chierici M (1998) A carbon budget for the Arctic Ocean. Glob Biogeochem Cycles 12:455–465

Andersson A, Hajdu S, Haecky P, Kuparinen J, Wikner J (1996) Succession and growth limitation of phytoplankton in the Gulf of Bothnia (Baltic Sea). Marine Biol 126:791–801

BACC Author Team (2008) Assessment of climate change for the Baltic Sea basin. Springer, Berlin, p 473

Bates NR (2006) Air–sea CO_2 fluxes and the continental shelf pump of carbon in the Chukchi Sea adjacent to the Arctic Ocean. J Geophys Res 111:C10013

Borges AV (2005) Do we have enough pieces of the jigsaw to integrate CO_2 fluxes in the coastal ocean? Estuaries 28:3–27

Borges AV, Delille B, Frankignoulle M (2005) Budgeting sinks and sources of CO_2 in the coastal ocean: diversity of ecosystems counts. Geophys Res Lett 32:L14601

Boutton TW (1991) Stable carbon isotopic ratios of natural materials. II. Atmospheric terrestrial, marine and freshwater environments. In: Coleman DC, Fry B (eds) Carbon isotope techniques. Academic Press, San Diego, pp 173–195

Bozec Y, Thomas H, Elkalay K, de Baar HJW (2005) The continental shelf pump for CO_2 in the North Sea-evidence from summer observation. Marine Chem 93:131–147

Burdige DJ, Homstead J (1994) Fluxes of dissolved organic carbon from Chesapeake Bay sediments. Geochimica et Cosmochimica Acta 58:3407–3424

Burdige DJ, Berelson WM, Coale KH, McManus J, Johnson KS (1999) Fluxes of dissolved organic carbon from California continental margin sediments. Geochimica et Cosmochimica Acta 63:1507–1515

Cai W-J, Dai MH, Wang YC (2006) Air–sea exchange of carbon dioxide in ocean margins: a province-based synthesis. Geophys Res Lett 33:L12603

Chen C-TA, Borges AV (2009) Reconciling opposing views on carbon cycling in the coastal ocean: Continental shelves as sinks and near-shore ecosystems as sources of atmospheric CO_2. Deep-Sea Res II 56:578–590

Chen C-TA, Liu K-K, Macdonald R (2003) Continental margin exchanges. In: Fasham MJR (ed) Ocean biogeochemistry. Springer, Berlin, pp 53–97

Cyberski J, Wróblewski A (2000) Riverine water inflows and the Baltic Sea water volume 1901–1990. Hydrol Earth Syst Sci 4:1–11

Dzierzbicka-Głowacka L, Kuliński K, Maciejewska A, Jakacki J, Pempkowiak J (2010) Particulate organic carbon in the southern Baltic Sea: numerical simulations and experimental data. Oceanologia 52(4):621–648

Emeis K-C, Struck U, Leipe T, Pollehne F, Kunzendorf H, Christiansen C (2000) Changes in the C, N, P burial rates in some Baltic Sea sediments over the last 150 years–relevance to P regeneration rates and the phosphorus cycle. Marine Geol 167:43–59

Emelyanov E (1995) Baltic Sea: Geology, geochemistry, paleoceanography, pollution. P.P. Shishov institute of oceanology RAS, Atlantic branch Baltic ecological institute of hydrosphere academy of natural sciences, RF, 115

Emerson SR, Hedges JI (2008) Chemical oceanography and the marine carbon cycle. Cambridge University Press, Cambridge, p 453

Frankignoulle M, Abril G, Borges A, Bourge I, Canon C, Delille B, Libert E, Théate J-M (1998) Carbon dioxide emission from European estuaries. Science 282:434–436

Fransson A, Chierici M, Anderson LG, Bussmann I, Kattner G, Jones EP, Swift JH (2001) The importance of shelf processes for the modification of chemical constituents in the waters of the Eurasian Arctic Ocean: implication for carbon fluxes. Cont Shelf Res 21:225–242

Fransson A, Chierici M, Nojiri Y (2006) Increased net CO_2 outgassing in the upwelling region of the southern Bering Sea in a period of variable marine climate between 1995 and 2001. J Geophys Res 111:C08008

Gerlach SA (1994) Oxygen conditions improve when the salinity in the Baltic Sea decreases. Marine Pollut Bull 28:413–416

Goyet C, Millero FJ, O'Sullivan DW, Eischeid G, McCue SJ, Bellerby RGJ (1998) Temporal variations of pCO_2 in surface sea water of the Arabian Sea in 1995. Deep-Sea Res I 45:609–623

Graham LP (2004) Climate change effects on river flow to the Baltic Sea. Ambio 33:235–241

Graham LP, Hagemann S, Jaun S, Beniston M (2007) On interpreting hydrological change from regional climate models. Clim Chang 81:97–122

Gudelis WK, Jemielianow JM (1982) Geologia morza bałtyckiego. Wydawnictwa Geologiczne, Warszawa, p 412

HELCOM (2004) The fourth Baltic Sea pollution load compilation (PLC-4). Baltic Sea Environ Proc 93:189

HELCOM (2006) Development of tools for assessment of eutrophication in the Baltic Sea. Baltic Sea Environ Proc 104:169

HELCOM (2009) Eutrophication in the Baltic Sea. Baltic Sea Environ Proc 115B, str. p 150

Henrichs SM, Reeburgh WS (1987) Anaerobic mineralization of marine sediment organic matter: rates and the role of anaerobic processes in the oceanic carbon economy. Geomicrobiol J 5:191–237

Hille S, Leipe T, Seifert T (2006) Spatial variability of recent sedimentation rates in the eastern Gotland Basin (Baltic Sea). Oceanologia 48:1–21

Holcombe BL, Keil RG, Devol AH (2001) Determination of pore-water dissolved organic carbon fluxes from Mexican margin sediments. Limnol Oceanogr 46:298–308

ICES (2008) Report of the ICES Advisory Committee, 2008. ICES Advice, 2008. Book 8, p 133

IPCC (2001) Climate change 2001. Synthesis report. A contribution of working groups I, II and III to the third assessment report of the intergovernmental panel on climate change. Cambridge University Press, Cambridge, p 398

IPCC (2007) Climate change 2007. Synthesis report. A contribution of working groups I, II and III to the fourth assessment report of the intergovernmental panel on climate change. Cambridge University Press, Cambridge, p 73

Jansson M (2001) Role of sediments in the nutrient dynamics of the Baltic. In: Wulff FV, Rahm LA, Larsson P (eds) A system analysis of the Baltic Sea. Springer, Berlin, pp 211–227

Jørgensen BB (1996) Case study—Aarhus Bay. In: Jørgensen BB, Richardson K (eds) Eutrophication in coastal marine ecosystems. Coastal and estuarine studies 52. AGU, Washington, pp 137–154

Kaltin S, Anderson LG (2005) Uptake of atmospheric carbon dioxide in Arctic shelf seas: evaluation of the relative importance of processes that influence pCO_2 in water transported over the Bering–Chukchi Sea shelf. Marine Chem 94:67–79

Karl DM, Bates NR, Emerson S, Harrison PJ, Jeandel C, Llinás O, Liu K-K, Marty J-C, Michaels AF, Miquel JC, Neuer S, Nojiri Y, Wong CS (2003) Temporal studies of biogeochemical processes determined from ocean time-series observations during the JGOFS era. In: Fasham MJR (ed) Ocean biogeochemistry. Springer-Verlag, Berlin, pp 239–267

Kuliński K (2010) Obieg węgla w Morzu Bałtyckim (Carbon cycling in the Baltic Sea). Ph.D. thesis, Institute of Oceanology PAS, p 134 (in Polish)

Kuliński K, Pempkowiak J (2008) Dissolved organic carbon in the southern Baltic Sea: Quantification of factors affecting its distribution. Estuarine Coast Shelf Sci 78:38–44

Kuliński K, Pempkowiak J (2011a) The carbon budget of the Baltic Sea. Biogeosciences. 8:3219–3230. doi:10.5194/bgd-8-3219-2011

Kuliński K, Pempkowiak J (2011b) Accumulation, mineralization and burial rates of organic carbon in the Baltic Sea sediments. Marine Chemistry, submitted

Kuliński K, She J, Pempkowiak J (2011) Short and medium term dynamics of the carbon exchange between the Baltic Sea and the North Sea. Cont Schelf Res 31:1611–1619

Kuss J, Roeder W, Wlost KP, DeGrandpre MD (2006) Time-series of surface water CO_2 and oxygen measurements on a platform in the central Arkona Sea (Baltic Sea): seasonality of uptake and release. Marine Chem 101:220–232

Lass H-U, Matthäus W (2008) General oceanography of the Baltic Sea. In: Feistel R, Nausch G, Wasmund N (eds) State and evolution of the Baltic Sea, 1952–2005. Wiley, Hoboken, pp 5–43

Łysiak-Pastuszak (2000) An assessment of nutrient conditions in the southern Baltic Sea between 1994–1998. Oceanologia 7:425–448

Macdonald RW, Naidu AS, Yunker MB, Gobeil C (2004) The Beaufort Sea: distribution, sources, fluxes and burial of organic carbon. In: Stein R, Macdonald RW (eds) The organic carbon cycle in the Arctic Ocean. Springer, Berlin, pp 177–193

Martens CS, Haddad RI, Chanton JP (1992) Organic matter accumulation, remineralization, and burial in an anoxic coastal sediment. In: Whelan JK, Farrington JW (eds) Organic matter: productivity, accumulation and preservation in recent and ancient sediments. Columbia University Press, New York, pp 82–98

Meier HEM (2006) Baltic Sea climate in the late twenty-first century: a dynamical downscaling approach using two global models and two emissions scenarios. Clim Dyn 27:39–68

Miller LA, Yager PL, Erickson KA, Amiel D, Bâcle J, Cochran JK, Garneau M-È, Gosselin M, Hirschberg DJ, Klein B, LeBlanc B, Miller WL (2002) Carbon distributions and fluxes in the North Sea, 1998 and 1999. Deep-Sea Res II 49:5151–5170

Naidu AS, Cooper LW, Grebmeier JM, Whitledge TE, Hameedi MJ (2004) The continental margin of the north Bering-Chukchi Sea: concentrations, sources, fluxes, accumulation and burial rates of organic carbon. In: Stein R, Macdonald RW (eds) The organic carbon cycle in the Arctic Ocean. Springer, Berlin, pp 193–203

Nausch G, Nehring D, Nagel K (2008) Nutrient concentrations, trends and their relation to eutrophication. In: Feistel R, Nausch G, Wasmund N (eds) State and evolution of the Baltic Sea, 1952–2005. Wiley, Hoboken, pp 337–393

Ohlson M (1990) Some aspects of a budget for total carbonate in the Baltic Sea. In: Proceedings of the 17th conference of the Baltic oceanographers, Norrköping, pp 78–83

Omstedt A, Elken J, Lehmann A, Piechura J (2004) Knowledge of the Baltic Sea physics gained during the BALTEX and related programmes. Prog Oceanogr 63:1–28

Omstedt A, Gustafsson E, Wesslander K (2009) Modelling the uptake and release of carbon dioxide in the Baltic Sea surface water. Cont Shelf Res 29:870–885

Papadimitriou S, Kennedy H, Bentaleb I, Thomas DN (2002) Dissolved organic carbon in sediments from eastern North Atlantic. Marine Chem 79:37–47

Pempkowiak J (1991) Enrichment factors of heavy metals in the Southern Baltic surface sediments dated with 210Pb and 137Cs. Environ Int 17:421–428

Pempkowiak J, Kupryszewski G (1980) The input of organic matter to the Baltic from the Vistula river. Oceanologia 12:80–98

Pempkowiak J, Bełdowski J, Kuliński K (2009) Zmiany globalne a zagrożenia dla, i ze strony Morza Bałtyckiego (Global changes and threat for and from the Baltic Sea). In: Gutry-Korycka M, Markowski T (eds) Zrównoważone warunki życia w zmieniającym się systemie klimatycznym Ziemi. Studia Tom CXXIV. Komitet Przestrzennego Zagospodarowania Kraju PAN, Warszawa, pp 137–149 (in Polish)

Prowe AEF, Thomas H, Pätsch J, Kühn W, Bozec Y, Schiettecatte L-S, Borges AV, de Baar HJW (2009) Mechanisms controlling the air-sea CO_2 flux in the North Sea. Cont Shelf Res 29:1801–1808

Rachold V, Eicken H, Gordeev VV, Grigoriev MN, Hubberten H-W, Lisitzin AP, Shevchenko VP, Schirrmeister L (2004) Modern terrigenous organic carbon input to the Arctic Ocean. In: Stein R, Macdonald RW (eds) The organic carbon cycle in the Arctic Ocean. Springer, Berlin, pp 33–55

Rullkötter J (2006) Organic matter: the driving force for early diagenesis. In: Schulz HD, Zabel M (eds) Marine Geochem. Springer, Berlin, pp 125–206

Staniszewski A (2005) Pochodzenie materii organicznej w osadach dennych Bałtyku Południowego (Origin of organic matter in bottom sediments of the south Baltic). Ph.D. Dissertation, Institute of Oceanology PAS, Sopot, p 185 (in Polish)

Stein R, Macdonald RW (2004) Organic carbon budget: Arctic Ocean vs. Global Ocean. In: Stein R, Macdonald RW (eds) The organic carbon cycle in the Arctic Ocean. Springer, Berlin, pp 315–363

Struck U, Emeis KC, Voss M, Christiansen C, Kunzendorf H (2000) Records of southern and central Baltic Sea eutrophication in $\delta^{13}C$ and $\delta^{15}N$ of sedimentary organic matter. Marine Geol 164:157–171

Suplinska M (2008) Sedimentation rates and dating of bottom sediments in the Southern Baltic Sea region. Nukleonika 53(Supplement 2):S105–S111

Szczepanska A, Zaborska A, Pempkowiak J (2009) Sediment accumulation rates in the Gotland Deep, Baltic Proper obtained by ^{210}Pb and ^{137}Cs methods. Annu Set Environ Prot 11(1):77–85

Szczepańska A, Maciejewska A, Zaborska A, Kuliński K, Pempkowiak J (2011) Distribution and provenience of organic carbon in the Baltic sediments. Geochronometria, accepted

Thomas H, Schneider B (1999) The seasonal cycle of carbon dioxide in Baltic Sea surface waters. J Marine Syst 22:53–67

Thomas H, Pempkowiak J, Wulff F, Nagel K (2003) Autotrophy, nitrogen accumulation and nitrogen limitation in the Baltic Sea: a paradox or a buffer for eutrophication. Geophys Res Lett 30:GL017937

Thomas H, Bozec Y, de Baar HJW, Elkalay K, Frankignoulle M, Schiettecatte L-S, Kattner G, Borges AV (2005) The carbon budget of the North Sea. Biogeosciences 2:87–96

Thomas H, Pempkowiak J, Wulff F, Nagel K (2010) The Baltic Sea. In: Liu K-K, Atkinson L, Quiñones RA, Talaue-McManus L (eds) Carbon and nutrient fluxes in continental margins. Springer, Berlin, pp 334–346

Voipio A (1981) The Baltic Sea. Elsevier Scientific Publishing Company, Amsterdam, p 418

Voss M, Larsen B, Leivuori M, Vallius H (2000) Stable isotope signals of eutrophication in Baltic Sea sediments. J Marine Syst 25:287–298

Voss M, Emeis K-C, Hille S (2005) Nitrogen cycle of the Baltic Sea from an isotopic perspective, Global biogeochemical cycles, 19:GB3001. doi:10.1029/2004GB002338

Wakita M, Watanabe YW, Watanabe S, Noriki S, Wakatsuchi M (2003) Oceanic uptake rate of anthropogenic CO_2 in a subpolar marginal sea: the Sea of Okhotsk. Geophys Res Lett 30:2252

Walsh JJ, Dieterle DA (1994) CO_2 cycling in the coastal ocean. I-A numerical analysis of the southeastern Bering Sea with applications to the Chuckchi Sea and the northern Gulf of Mexico. Prog Oceanogr 34:335–392

Wasmund N, Siegel H (2008) Phytoplankton. In: Feistel R, Nausch G, Wasmund N (eds) State and evolution of the Baltic Sea, 1952–2005. Wiley, Hoboken, pp 441–481

Wasmund N, Uhlig S (2003) Phytoplankton trends in the Baltic Sea. J Marine Syst 60:177–186

Wasmund N, Andrushaitis A, Łysiak-Pastuszak E, Müller-Karulis B, Nausch G, Neumann T, Ojaveer H, Olenina I, Postel L, Witek Z (2001) Trophic status of the south-eastern Baltic Sea: a comparison of coastal and open areas. Estuarine Coastal Shelf Sci 53:849–864

Widrowski H, Pempkowiak J (1986) The history of surface sediments in the Southern Baltic. In: Proceedings of the 15th conference Baltic oceanographers, Marine Pollution Laboratory, Copenhagen, pp 656–671

Wulff F, Rahm L, Hallin A-K, Sandberg J (2001) A nutrient budget model of the Baltic Sea. In: Wulff FV, Rahm LA, Larsson P (eds) A system analysis of the Baltic Sea. Springer, Berlin, pp 354–372

Zaborska A (2007) Benthic sedimentary processes and organic matter burial in the northwestern Barents Sea. Ph.D. Disseration, Institute of Oceanology, Sopot, p 181

Zaborska A, Carrol J, Papucci C, Torricelli L, Carrol M, Walkusz-Miotk J, Pempkowiak J (2008) Recent sediment accumulation rates for the Western margin of the Barents Sea. Deep-Sea Res II 55:2361–2371

Zhai W, Dai M, Cai W-J, Wang Y, Wang Z (2005) High partial pressure of CO_2 and its maintaining mechanism in a subtropical estuary: the Pearl River estuary, China. Marine Chem 93:21–32

Appendix
Methods Used for Quantification of Selected Carbon Fluxes in the Baltic Sea

A.1 Carbon Export to and Import from the North Sea

A.1.1. The Study Area

The Baltic Sea is connected to the North Sea via Skagerrak, Kattegat, as well as the narrow and shallow Danish Straits: Great Belt, Little Belt and Oresund (Fig. A.1).

Additional barriers for the water exchange are the Darss and Drogden Sills located at the entrance to the Baltic Sea. This specific sea bottom geomorphology contributes to limitation in the highly saline and well oxygenated North Sea water inflows. Moreover, the complexity of the hydrological conditions in this transition area is intensified due to eddy occurrence. Döös et al. (2004) estimated that only 6 and 32% of the inflowing North Sea water via Great Belt and Oresund, respectively, remain in the Baltic for a period longer than one year. To eliminate these high fluctuations from the carbon exchange calculations, three latitudinal transects in the Danish Straits were selected, where DMI-BSHcmod—a hydrostatic 3D circulation model (Sect. 1.2; She et al. 2006, 2007)—was engaged (Fig. A.1). Since water flowing through the Danish Straits is most often a mixture of the North Sea and the Baltic Sea water masses, it was desirable to distinguish between them in the carbon exchange calculations. For this reason two "end member" stations were selected: 'MB'—located in the vicinity of the Arkona Deep (55°01′N; 14°04′E), characteristic for the "pure" Baltic and 'MP'—located in the eastern North Sea (57°21′N; 6°37′E) characteristic of the North Sea water masses (Fig. A.1).

A.1.2. Carbon Fluxes Quantification

Quantification of carbon fluxes between the Baltic Sea and the North Sea required simultaneous research on both: quantifications of the water masses exchange and

K. Kuliński and J. Pempkowiak, *Carbon Cycling in the Baltic Sea*, Geoplanet: Earth and Planetary Sciences, DOI: 10.1007/978-3-642-19388-0,

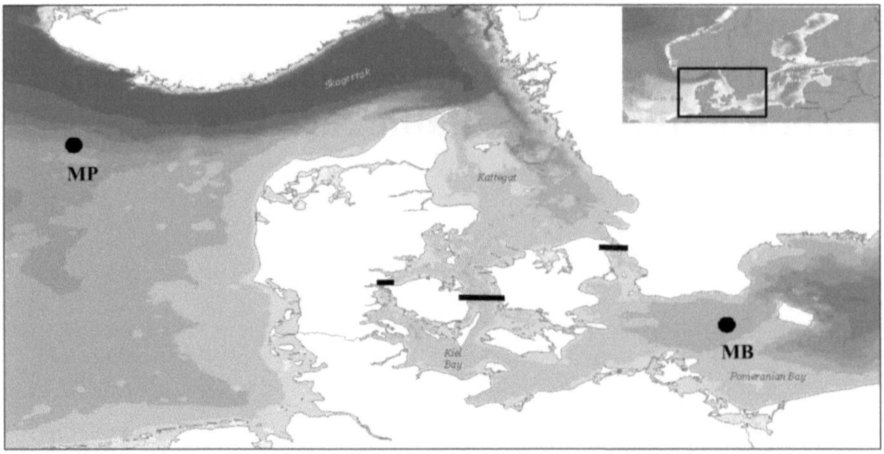

Fig. A.1 Localizations of the MP and MB end member stations and latitudinal transects in the Danish Straits (*black* lines) (Kuliński et al. 2011)

the carbon concentrations of the studied water masses. Methodologies used in both these ventures are described separately.

Hydrology. The hourly transports through the sections (Fig. A.1), for a 4-year long period, are estimated by using a three-dimensional (3D) ocean circulation model—DMI-BSHcmod, which is a Danish Meteorological Institute (DMI) operational model. The model was originally developed by BSH, the Bundesamt für Seeschifffahrt und Hydrographie (Federal Maritime and Hydrographic Agency) in Germany. It has been further developed and calibrated for operational storm surge forecast in DMI since 1999 (She et al. 2006). The model is based on a simple equation for momentum conservation in a spherical coordinate, a state equation and hydro-static approximation in the vertical direction. The turbulence vertical mixing scheme is based on the k-ω turbulence model extended for buoyancy affected by geophysical flows (Umlauf et al. 2003), but with a new set of coefficients. Through parameterizations it takes into account breaking surface waves. The surface heat flux is parameterized using bulk quantities of both atmosphere and sea or sea ice, respectively.

The model has three nested layers to resolve the exchange between the Baltic Sea, the North Sea and the North Atlantic Shelf Sea. The outer layer model is a 2D surge model, covering the Shelf Sea, to provide the surge boundary conditions for a one-way nested 3D Baltic–North Sea model. Both models have a horizontal resolution of 6 nautical miles. In order to resolve the narrow Danish Straits linking the Baltic and North Sea, a third layer high resolution (1 nautical mile) 3D ocean model is used, which covers only the Baltic–North Sea transition region and has a two-way nesting with the 3D Baltic–North Sea model. All three models are discrete horizontally in regular spherical grids and vertically in the z levels. The 3D models have in total 50 vertical layers. The fine grid model of the Danish inner waters has nominally 1 m vertical resolution, however top and bottom layers may differ in their thickness due to the tides and different reservoir depths. The model is forced by hourly meteorological parameters based on DMI's operational weather model HIRLAM (High Resolution Limited Area Model), with a horizontal resolution of about 15 km. River runoff is daily averaged based on data derived from river measurements for 5 German rivers, hydrological simulations for 42 Baltic rivers and climatology values for the remaining rivers. The lateral boundary for the Baltic–North Sea model includes wind-induced sea level from the 2D Shelf Sea model, tidal water level from the 14 tidal constituents and monthly mean water temperature and salinity profiles along the boundary.

For the purpose of carbon exchange quantification, longitudinal components of the water currents and salinities (S_{CD}) of the water flowing through the selected transects in the Danish Straits were determined. Calculated water transport covered the period between 1 June 2002 and 31 May 2006. Assuming that the North Sea and the Baltic Sea water masses influence salinity in the Danish Straits, their contributions in the bulk of the flowing water were separated based on the salinity of the end members (EM) approach described by the following equations:

$$x + y = 1 \quad \text{(A.1)}$$

$$xS_B + yS_P = S_{CD} \quad \text{(A.2)}$$

where: x and y are the contributions of, respectively, the Baltic Sea and the North Sea water masses flowing through the investigated transects, S_B, S_P are the salinities of, respectively, the Baltic Sea and the North Sea water masses at the EM stations and S_{CD} is the modelled salinity of water flowing through the Danish Straits. S_B values in the considered period were obtained from the Baltic Sea Research Institute in Warnemuende that conducts CTD monitoring measurements on behalf of the Federal Maritime and Hydrographic Agency (BSH) on the buoy located in the Arkona Basin at MARNET Monitoring Station (close to the MB station). Dynamics of the S_p was adopted from Bozec et al. (2006) and Prowe et al. (2009). In both cases salinity end member data were averaged with the seasonal resolution. The mean salinity at the MB station during the investigated period of time (2002–2006) amounts to 7.7 ± 0.2 PSU whilst at the MP station it amounts to 34.2 ± 0.5 PSU.

The determined volumes of the Baltic Sea water and the North Sea water were summed up in the daily periods of time, separately for water inflows and outflows. In this way four different water flow values were calculated, namely: (1) the Baltic Sea water exported to the North Sea, (2) the Baltic Sea water returning back from the North Sea to the Baltic, (3) the North Sea water flowing to the Baltic and (4) the North Sea water returning back from the Baltic to the North Sea.

Carbon concentrations. For the purpose of carbon exchange quantification between the Baltic Sea and the North Sea, dissolved inorganic carbon (DIC) and dissolved organic carbon (DOC) surface concentrations seasonality at MB and MP were considered. When no data were available directly for the MB and MP stations, the data from as close as possible locations that characterize a typical Baltic Sea and North Sea water masses were used. Particulate carbon species were not included in the calculations due to their relatively low contributions to the total carbon (Nausch et al. 2008; Thomas et al. 2010).

The seasonality of DIC concentrations at MB and MP stations is adopted from Thomas and Schneider (1999) and Prowe et al. (2009), respectively. In both cases, DIC concentrations data are based on the model estimations validated by the experimental data sets. Uncertainty of these results was set as suggested by the authors. Additionally, surface seawater samples for the DIC analysis were collected during the research cruises of r/v Oceania during 2006–2009. DIC concentrations at the MB and MP stations were interpolated to obtain daily resolution of carbon concentrations data fitting to the hydrological data resolution.

There is lack of reliable literature data related to the DOC seasonal variations for the investigated areas. Therefore, weekly DOC measurements were carried out within the Gulf of Gdansk. In order to eliminate short-term DOC concentrations variability, the moving average (three weeks average period) was used. The results so obtained were extrapolated to the investigated EM stations based on the DOC measurements carried out at MB and MP stations, as well as on the results published by Suratman et al. (2009). This approach was adopted since DOC concentrations variability in seawater is greatly influenced by the biological activity (Thomas and Schneider 1999; Carlson 2002; Kuliński and Pempkowiak 2008; Prowe et al. 2009). Uncertainty of the DOC concentrations at MB and MP stations was set as Relative Standard Deviation (RSD) of the analytical method used. These were calculated from the Standard Deviation (SD) of moving averages.

DIC analysis. DIC samples were collected directly from a bathometer to the 40 ml glass ampoules, each poisoned with 150 μl of saturated $HgCl_2$ solution. Analysis was carried out in a 'HyPerTOC' analyzer (Thermo Electron Corp., the Netherlands), using modified method based on sample acidification and detection of arising CO_2 in the non-dispersive infrared (NDIR) detector (Kaltin et al. 2005). DIC concentrations in the analyzed samples were calculated from the calibration line obtained using Na_2CO_3 standard solutions. Each sample was analyzed in triplicate and the precision assessed as RSD was better than 0.8%.

DOC analysis. DOC samples were collected using a bathometer. Immediately after collection, the water was passed through a 0.2 μm pre-cleaned membrane filter. A total of 40 ml aliquots of filtrate were acidified with 150 μl of concentrated HCl and stored in the dark at 5°C until further laboratory analysis. These were carried out in a 'HyPerTOC' analyser using UV/persulphate oxidation method and NDIR detection (Sharp 2002; Kuliński and Pempkowiak 2008). In order to remove inorganic carbon, DOC samples were purged before analysis with the compressed, CO_2 free air. DOC concentrations were calculated using the standard, potassium hydrogen phthalate addition method. Quality control for DOC data was provided by analyzing standard sea water obtained from CRM's (supplied by the Hansell Laboratory, University of Miami), used with each sample sequence. The recovery was not worse than 94% and the precision as described by the RSD of a sample single triplicate analysis was not worse than 3%.

A.2 Deposition and Burial of Carbon in the Baltic Sediments

A.2.1. Rationale

It is believed that the oceans constitute a great sink for the anthropogenic CO_2, thus reducing its negative impact on the global climate. Among the processes that govern the atmospheric CO_2 uptake by the oceans, biological pump plays a key role. Biological pump is forced by the photosynthesis during which CO_2 is turned into C_{org}, incorporated into biomass and transported along the food chain as an organic matter. A significant part of the dead organic matter is mineralized in the water column and on the sediment surface. However, organic carbon (C_{org}) burial in sediments is also observed. Globally, it is estimated that from 0.01 to 1% of the primary production is buried in the marine sediments and excluded from the earth crust carbon cycle (Rullkötter 2006).

Since the shelf seas are responsible for production of 20% of newly produced organic matter, and up to 80% of the total organic matter exported to the sea bottom in the global ocean (Borges 2005), they are believed to play a significant role in the course of anthropogenic CO_2 uptake (Chen and Borges 2009). The role of the shelf seas becomes more important when their relatively small sea surface is considered. The importance of global shelf is strengthened by environmental conditions that occur both in the water column and in the bottom sediments. High primary production in the upper, eutrophic zone causes high organic matter deposition to the surface sediments (Rullkötter 2006). The latter, together with often limited ventilation of the near-bottom water layers results in hypoxic or even anoxic conditions in the sediments. As a consequence both the lower oxidation potential and the lower microbiological activity enhance organic matter burial in the sediments (HELCOM 2009).

These facts were the motivation for the presented study which aimed at quantification of the C_{org} burial in the bottom sediments of the Baltic Sea. For the purpose of this study the C_{org} burial rate quantifications base on the measurements of both accumulation rates of the sedimentary matter and C_{org} concentrations in the matter (Carman et al. 1996; Emeis et al. 2000; Müller 2002).

However, this approach does not take into consideration of the long-term mineralization of organic matter. This is especially important since the organic matter accumulation in sediments takes place mostly in the deep, depositional areas due to the near-bottom water currents that flush the settled matter out from the shallow areas (Jönsson et al. 2005). It causes the high accumulation of the organic matter in depositional areas. Together with the limited ventilation of the near-bottom water this results in the reduction of the red-ox potential. This causes an extension of the time required for the mineralization of the labile organic matter (Schulz 2006). Thus, additional goals of the study were set: (1) to quantify the influence of the long-term mineralization on the C_{org} burial in the Baltic sediments and (2) to determine the dynamics of the C_{org} mineralization in the subsurface levels of the sediments column.

A.2.2 Carbon Deposition to Sediments

For the purpose of the C_{org} deposition to the Baltic Sea sediments (excluding Kattegat) a number of the literature data were used (Błaszczyszyn 1982a, b; Carman et al. 1996; Emeis et al. 2000; Algesten et al. 2006; Christoffersen et al. 2007). Since the bulk of the settled particulate organic matter (POM), due to the hydrodynamical conditions, is finally deposited in the deep, depositional areas of the Baltic Sea, namely: the Arkona Basin, the Bornholm Deep, the Gdansk Deep, the Gotland Deep, the Gulf of Finland, the Gulf of Riga and the Gulf of Bothnia (Błaszczyszyn 1982a; Jönsson et al. 2005; Schmidt et al. 2008), the C_{org} load to the sediments was calculated as the product of the depositional areas surface (Błaszczyszyn 1982a; Kuliński 2010), and the C_{org} accumulation rates there (Emeis et al. 2000; Christoffersen et al. 2007; Kuliński 2010). For the Gulf of Bothnia and the Gulf of Riga, the amounts of C_{org} exported to the sediments were adopted from the papers by Algesten et al. (2006) and Carman et al. (1996), respectively. Due to the lack of C_{org} accumulation rates from the Gulf of Finland, the results from the Gotland Deep were considered. Kattegat was not investigated in this study.

Generally, C_{org} accumulation rates are calculated as a product of sedimentary material deposition rates and C_{org} concentrations in sedimentary material. Hence the results obtained using this method account for the portion of C_{org} which was not mineralized during the sedimentation in the water column and which was not mineralized at the sediment surface. According to this definition, C_{org} accumulation rates can be regarded as organic carbon flux to sediments (or alternatively—organic carbon deposition to sediments). However, the organic carbon flux to sediments (C_{org} sedimentation rate, C_{org} deposition rate) is often mistaken with carbon burial (e.g., Carman et al. 1996; Emeis et al. 2000; Müller 2002; Zeldis et al. 2010). The carbon burial rates calculated by this method are overestimated due to the long-term organic matter mineralization that occurs in the subsurface layers of the sediments column, often for tens of years. Hence, loads of C_{org} deposited presently to the bottom sediments of the Baltic Sea depositional areas were corrected, in this study, by introducing carbon loss from sediments due to the sedimentary organic matter mineralization. We call it the carbon return flux.

A.2.3 Carbon Return Flux

Since both the dissolved inorganic and the dissolved organic carbon species (DIC and DOC, respectively), are a product of organic matter mineralization and decomposition in the sediments (Rullkötter 2006; Emerson and Hedges 2008), the return carbon flux related to the long-term mineralization of organic

Table A.1 Coordinates, dates of sampling and water depths at the sampling stations (Kuliński and Pempkowiak 2011b)

Station	Sampling date	Latitude	Longitude	Depth (m)
MB	14.05.2007	55°01′N	14°04′E	45
B1	26.04.2008	55°12′N	15°30′E	70
P1	01.05.2008	54°50′N	19°20′E	107
BY15	27.01.2009	57°19′N	20°03′E	243
ZF	21.01.2009	65°23′N	23°27′E	91

matter was calculated using the Fick's First Law of diffusion (Ullman and Aller 1982):

$$J = -\phi \cdot D_{\text{sed}} \cdot \frac{\Delta C}{\Delta x} \tag{A.3}$$

where: J is the diffusion flux of carbon species (solutes) (μg cm^{-2} s^{-1}), ϕ is the porosity of the sediment (%), D_{sed} is the sediment diffusion coefficient of a solute (cm^2 s^{-1}) and $\frac{\Delta c}{\Delta x}$ is the solute concentrations gradient in upper layer of the sediment core $\left(\frac{\mu\text{g cm}^{-3}}{\text{cm}}\right)$ For the calculations done in this study, the following D_{sed} values were adopted: 6.32·10^{-6} cm^2 s^{-1} (Martin and McCorkle 1993) and 1.22·10^{-6} cm^2 s^{-1} (Holcombe et al. 2001) for DIC and DOC, respectively.

For the purpose of the DIC and DOC concentrations gradient calculations in the pore waters, five sediment cores were collected using GEMAX corer from the crucial Baltic Sea depositional areas in the period from 2007 to 2009 (Table A.1; Fig. A.2).

Immediately after collecting, the sediment cores were sliced into 1 cm thick layers and frozen in the Ar atmosphere (−20°C). In the laboratory, thawed samples were centrifuged (15,000G). Pore water was decanted to two 20 ml glass vials and conserved with 75 μl of saturated $HgCl_2$ (DIC samples) and 75 μl of concentrated HCl (DOC samples).

Analyses of both DIC and DOC samples were done in a 'HyPerTOC' analyser (Thermo Electron Corp., the Netherlands). A modified method based on the samples acidification and detection of purged CO_2 in the non-dispersive infra red (NDIR) detector (Kaltin et al. 2005) was used for the DIC measurements. The DIC concentrations in the analyzed samples were calculated from the calibration line obtained using Na_2CO_3 standard solutions. Each sample was analyzed in triplicate. The obtained precision, assessed as RSD, was better than 1.7%. The DOC analyses were performed using UV/persulphate oxidation method and NDIR detection (Sharp 2002; Kuliński and Pempkowiak 2008). Each DOC sample was purged before analysis with the synthetic, CO_2-free air in order to remove inorganic carbon. DOC concentrations in the samples were measured using the standard—potassium hydrogen phthalate—addition method. Quality control of DOC results was carried out with analysis of CRM's supplied by the Hansell Laboratory, University of Miami, with each sample's sequence. The recovery was not worse

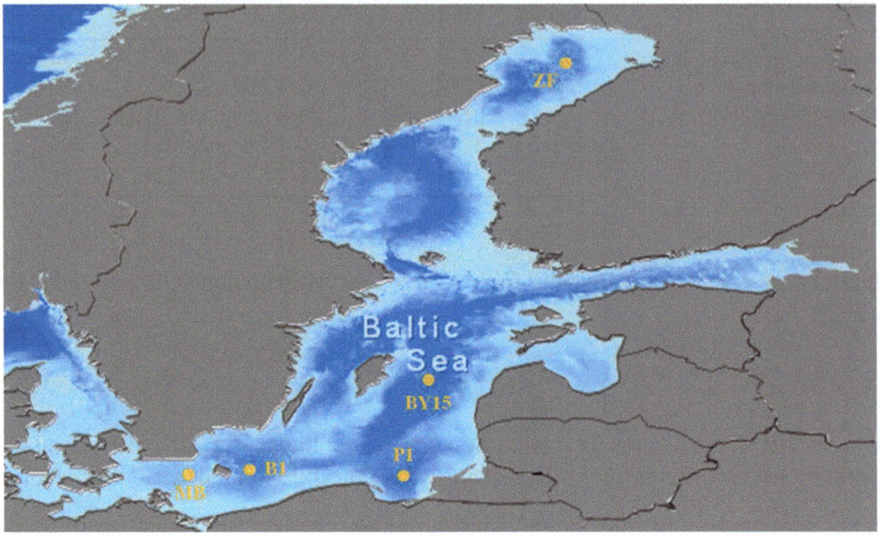

Fig. A.2 Localization of sediments cores sampling stations (*yellow*) (Pempkowiak et al. 2011b)

than 95% and the precision, as described by the RSD of a single sample triplicate analysis, was not worse than 2.8%.

Due to the lack of sediment cores sampled in the Gulf of Finland and Gulf of Riga, the diffusion fluxes of DIC and DOC were adopted from the Gotland Deep for these areas.

A.2.4 Dynamics of the Organic Matter Long-Term Mineralization

The long-term mineralization dynamics of the sedimentary organic matter was investigated during the laboratory experiment. A sediment core was collected, using a GEMAX corer, at the P1 station during the r/v Oceania expedition (Table A.1; Fig. A.1). The core was subdivided into four cores of smaller diameter by means of a tube divided into four smaller tubes. The sediment cores so-collected, together with the overlaying near-bottom water, were closed tightly in the Ar atmosphere and stored in dark at 4°C till the start of the laboratory experiment. Immediately after the delivery of the sediment cores to the laboratory, one of them was sliced into 1 cm layers, freeze-dried and homogenized. Some 200 mg of dry sediment of the so prepared samples were used for the dating procedure. This was done using the ^{210}Pb method (Joshi and Shukla 1991; Pempkowiak 1991). The remaining bulks of samples were acidified in the fumes of concentrated HCl for 24 h in order to remove inorganic carbon species. Afterwards, the samples were again dried and homogenized. In these prepared samples, concentrations of C_{org}, total nitrogen and total phosphorus were measured in order to investigate the dynamics of the organic matter mineralization.

Samples for C_{org} and nitrogen analysis were weighed into silver capsules (accuracy of 0.000001 g), acidified and analyzed simultaneously in an Elemental Analyzer Flash EA 1112 Series combined with the Isotopic Ratio Mass Spectrometer IRMS Delta V Advantage (Thermo Electron Corp., Germany). Quality control of C and N concentrations analysis was carried out with standard materials supplied by the Thermo Electron Corp. The methodology ensured satisfactory accuracy (average recovery of 99.0% for C and 98.3% for N) and precision characterized by the relative standard deviation (RSD) better than 1.7 and 1.9% (n = 5) for C and N, respectively.

Phosphorus concentrations in the sediment samples were measured colorimetrically (Golachowska 1997). In the glass tubes, 100 mg of the dry sediment was mineralized in the temperature gradient (100–300°C) with 1.6 ml of the concentrated acids mixture ($HClO_4$ and H_2SO_4). The mineralized samples were diluted with Milli Q water and filtered through MN GF5 glass fiber filters. Filtrates were transferred to the 100 ml flasks and neutralized with NaOH in the presence of phenolphthalein. Then flasks were filled up to 100 ml with Milli Q water and 1 ml $(NH_4)_6Mo_7O_{24}$ and 0.5 ml $SnCl_2$ solutions. After 10–12 min phosphorus concentrations were measured by the colorimetric method on the spectrofluorometer GENios. For each of samples series, RSD was established from a single sample triplicate analysis. It was not worse than 2.4%.

The three remaining cores were incubated in dark at 17°C. After 141, 262 and 370 days, successive cores were analyzed according to the above-mentioned procedures. Based on the obtained results, annual losses of C_{org} were calculated at the subsequent levels of sediment core. The ratios C/N/P enabled to define the impact of the mineralization process on the organic matter concentration. Simultaneously, the dynamics of the organic matter mineralization was described by the velocities of C_{org} loss calculated according to the first-order kinetics reaction:

$$v = \frac{dC}{dt} = k \cdot C_0 \tag{A.4}$$

where: v is the mineralization velocity (mg C g^{-1} day^{-1}), $\frac{dc}{dt}$ is the C_{org} concentration change in time t (mg C g^{-1} day^{-1}), k is the mineralization constant (day^{-1}) and C_0 is the initial C_{org} concentration (mg C g^{-1}). It was assumed that temperature increase of every 10°C from the in situ temperature causes double increase of k, which was calculated from the integrated form of the equation A.4:

$$k = -\ln \sqrt[t]{\frac{C_t}{C_0}} \tag{A.5}$$

where: C_t is the C_{org} concentration after time t (mg C g^{-1}) and C_0 is the initial C_{org} concentration in time t_0 (mg C g^{-1}).

A.2.5 Organic Carbon Burial

Organic carbon burial rate was calculated as a difference between the deposition and mineralization rates of organic carbon, as described in Sect. A.2.2.

A.3 Organic Matter Deposition and Origin in the Baltic Sea Sediments

Experimental. The southern part of the Baltic Proper has been selected for the study on the deposition rates and provenience of sedimentary organic matter. The sampling stations cover the following areas of the Baltic Proper: the Gotland Deep and the Gdansk Deep. The deeps are known as accumulation areas of fine sedimentary material of both autochtonous and allochtonous origin (Gudelis and Jemielianov 1982).

For the purpose of the deposition and provenience evaluation, sediment cores were collected using a GEMAX corer during the r/v Oceania cruises at 6 stations in the Baltic Sea (Table A.2, Fig. A.3). Immediately after collecting, the cores were sliced into 10 mm thick layers and the obtained sediment samples were deep frozen. After thawing in the laboratory, the sediment samples were freeze-dried and homogenized.

Sediment accumulation rates (SAR in cm yr^{-1}) were determined from profiles of excess ^{210}Pb ($^{210}Pb_{supp}$ was subtracted from total ^{210}Pb) versus porosity-corrected depth in sediments. Determination of ^{210}Pb activity concentration was performed assuming secular equilibrium between ^{210}Pb and ^{210}Po, according to the method developed by Flynn (1968) and adopted by Widrowski and Pempkowiak (1986). In short: 100 mg of dry sediment was spiked with known activity of ^{209}Po, then the sample was digested with 2 cm^3 of perchloric acid followed by 4 cm^3 of fluoric acid. The obtained solution was transferred, with 6 M HCl, to a teflon vessel and evaporated to dryness. The dry residue was dissolved in 50 ml of 0.5 mol dm^{-3} hydrochloric acid and both ^{209}Po and ^{210}Po were deposited onto silver disks (90°C, 4 h, continuous stirring). The disks were counted for 20 h in a multi-channel analyzer (Canberra).

Sediment accumulation rates were determined basing on an exponential decrease of $^{210}Pb_{ex}$ activity with sediment depth. In the case of surface sediment layers mixing, sediment accumulation rates were calculated below the mixed zone. The following formulas have been used for calculating sediment accumulation rates:

$$A_t = A_0 \, e^{-\lambda t} \qquad (A.6)$$

where: A_t is the ^{210}Pb activity at time t, A_0 is the activity at time 0, λ is the radionuclide decay constant (for ^{210}Pb, λ = 0,031). When t is replaced by $t = x/v$

Table A.2 Location of cores sampling sites and the collected sediments porosity. (Szczepańska et al. 2011)

Station	Latitude	Longitude	Depth (m)	Core length (cm)	Porosity (%)
I	57°54′ N	17°18′ E	80	25	0.76–0.89
II	56°37′ N	19°20′ E	135	28	0.81–0.98
III	57°19′ N	19°54′ E	230	23	0.76–0.97
IV	56°19′ N	18°36′ E	85	23	0.70–0.93
V	54°37′ N	19°30′ E	55	42	0.81–0.95
VI	56°03′ N	16°02′ E	147	35	0.86–0.94

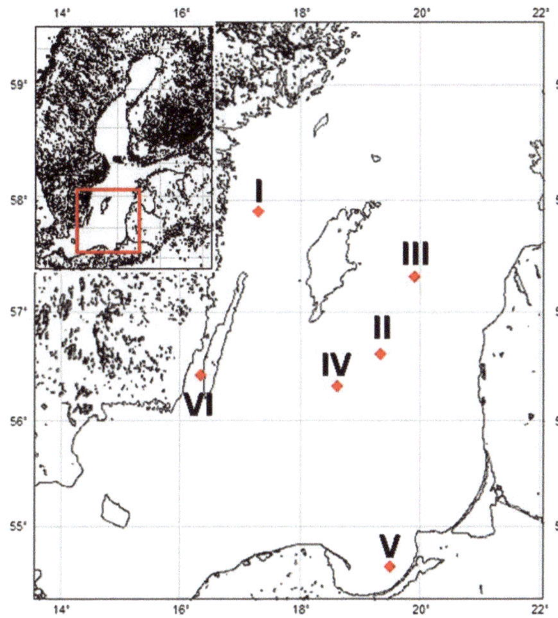

Fig. A.3 Location of sampling stations (Szczepańska et al. 2011)

(x—depth of a given sediment layer, v—sedimentation rate) the above formula can be rewritten as:

$$\ln A^{210}Pb_{ex}(x) = \ln A^{210}Pb_{ex}(0) - (\lambda/v)x \tag{A.7}$$

where: $A^{210}Pb_{ex}(x)$ is the activity at layer x, $A^{210}Pb_{ex}(0)$ is the activity at surface (layer 0), λ is the decay constant and v is the sediment accumulation rate. The constant rate of supply (CRS) model was used throughout the calculations (Robins 1978; Pempkowiak and Widrowski 1982; Zaborska et al. 2007).

$$A_x = A_0\, e^{-\lambda t} \tag{A.8}$$

$$x/v = 1/\lambda \ln \sum A_{cx} \Big/ \ln \sum A_{c0} \tag{A.9}$$

where: ΣA_{cx} is the cumulative residual $^{210}Pb_{ex}$ activity beneath sediments of depth x, ΣA_{c0} is the total $^{210}Pb_{ex}$ activity in the sediment column, x is the cumulative depth (cm), v is the sediment accumulation rate (cm yr^{-1}), t is the age of sediment (years).

Sediment mass accumulation rates (MAR in g m^{-2} yr^{-1}) were calculated using the $^{210}Pb_{ex}$ after transforming sediment depth to mass depth (Robins 1978; Zaborska et al. 2008). The cumulative dry mass depth was estimated according to the formula:

$$D_D = h \cdot D_s \cdot (1 - P) \tag{A.10}$$

$$\sum D_D = \sum (h_{i+1} - h_1) \cdot (1 - P_i) \cdot D_i \tag{A.11}$$

where: D_D is dry mass depth, h is the layer thickness, D_s is dry sediment density and P is the layer porosity.

To validate the obtained sediment accumulation rates, profiles of the man-made radionuclide, ^{137}Cs, were used. Radionuclide ^{137}Cs has been introduced into the environment as a consequence of nuclear weapons tests in the period from 1950 to 1976 (Abril 2003). The significant quantities of ^{137}Cs were introduced also as a result of the Chernobyl accident in 1986 (Kankaanpaa et al. 1997; Lima 2004; Suplińska 2002). The ^{137}Cs activity was determined using gamma spectrometry. Calibration of the counter was performed based on the reference IAEA materials. The counting time was at least 24 h.

Organic carbon and total nitrogen concentrations, and stable carbon isotopes composition ($\delta^{13}C$) were measured in an Elemental Analyzer Flash EA 1112 Series combined with the Isotopic Ratio Mass Spectrometer IRMS Delta V Advantage (Thermo Electron Corp., Germany). The reference gases were pure N_2 and CO_2 gas calibrated against IAEA standards: CO-8 and USGS40 for $\delta^{13}C$ and N-1 and USGS40 for $\delta^{15}N$. The standard deviation was less than 0.08‰ and 0.11‰ for $\delta^{13}C$ and $\delta^{15}N$, respectively. Quality control of C and N concentrations was carried out with standard materials supplied by the Thermo Electron Corp. The methodology is good to achieve a satisfactory accuracy and precision (average recovery: 99.1% for C and 98.5% for N). Before analyzing, carbonates were removed from the samples by acidification (HCl_{conc}) for 24 hours.

Organic carbon and total nitrogen accumulation rates were calculated as a product of mass sediment accumulation rates (g m^{-2} yr^{-1}) and organic carbon/ total nitrogen concentrations. Only the surface layers, above the sediment mixed depth, organic carbon concentrations were used, to get a newly deposited organic carbon flux to sediments.

References

Abril JM (2003) Constraints on the use of ^{137}Cs as a time-marker to support CRS and SIT chronologies. Environ Pollut 129:31–37

Algesten G, Brydsten L, Jonsson P, Kortelainen P, Löfgren S, Rahm L, Räike A, Sobek S, Tranvik L, Wikner J, Jansson M (2006) Organic carbon budget for the Gulf of Bothnia. J Marine Syst 63:155–161

Błaszczyszyn A (1982a) Środowisko sedymentacyjne Morza Bałtyckiego (Sedimentational environment of the Baltic Sea). In: Gudelis WK, Jemielianow JM (eds) Geologia Morza Bałtyckiego. Wydawnictwa Geologiczne, Warszawa, pp 131–141 (in Polish)

Błaszczyszyn A (1982b), Główne składniki chemiczne osadów dennych Morza Bałtyckiego (Main chemical components of bottom sediments in the Baltic Sea). In: Gudelis WK, Jemielianow JM (eds) Geologia Morza Bałtyckiego. Wydawnictwa Geologiczne, Warszawa, pp 257–289 (in Polish)

Borges AV (2005) Do we have enough pieces of the Jigsaw to integrate CO_2 fluxes in the Coastal ocean? Estuaries 28:3–27

Bozec Y, Thomas H, Schiettecatte L-S, Borges AV, Elkalay K, de Baar HJW (2006) Assessment of the processes controlling seasonal variations of dissolved inorganic carbon in the North sea. Limnol Oceanogr 51:2746–2762

Carlson CA (2002) Production and removal processes. In: Hansell DA, Carlson CA (eds) Biogeochemistry of marine dissolved organic matter. Elsevier Science, San Diego, pp 91–151

Carman R, Aigars J, Larsen B (1996) Carbon and nutrient geochemistry of the surface sediments of the Gulf of Riga, Baltic sea. Marine Geol 134:57–76

Chen C-TA, Borges AV (2009) Reconciling opposing views on carbon cycling in the coastal ocean: Continental shelves as sinks and near-shore ecosystems as sources of atmospheric CO_2. Deep-Sea Res II 56:578–590

Christoffersen PL, Christiansen C, Jensen JB, Leipe T, Hille S (2007) Depositional conditions and organic matter distribution in the Bornholm Basin, Baltic sea. Geo-Marine Lett 27:325–338

Döös K, Meier HEM, Döscher, R (2004) The Baltic haline conveyor belt or the overturning circulation and mixing in the Baltic. Ambio 33:261–266

Emeis K-C, Struck U, Leipe T, Pollehne F, Kunzendorf H, Christiansen C (2000) Changes in the C, N, P burial rates in some Baltic sea sediments over the last 150 years–relevance to P regeneration rates and the phosphorus cycle. Marine Geol 167:43–59

Emerson SR, Hedges, JI (2008) Chemical oceanography and the marine carbon cycle. Cambridge University Press, Cambridge, p 453

Flynn WW (1968) The determination of ^{210}Po in environmental materials. Analytica Chimica Acta 43:221–227

Gudelis WK, Jemielianow JM (1982) Geologia Morza Bałtyckiego. Wydawnictwa Geologiczne, Warszawa, p 412

HELCOM (2009) Eutrophication in the Baltic sea. Baltic sea environment proceedings 115B, str. 150 pp

Holcombe BL, Keil RG, Devol AH (2001) Determination of pore-water dissolved organic carbon fluxes from Mexican margin sediments. Limnol Oceanogr 46:298–308

Jönsson A, Danielsson Å, Rahm L (2005) Bottom type distribution based on wave friction velocity in the Baltic sea. Cont Shelf Res 25:419–435

Joshi SR, Shukla BS (1991) At initial derivation of formulation Pb-210 dating of sediments. J Radioanal Nucl Chem 148:73–79

Kaltin S, Anderson LG (2005) Uptake of atmospheric carbon dioxide in Arctic shelf seas: evaluation of the relative importance of processes that influence pCO_2 in water transported over the Bering–Chukchi sea shelf. Marine Chem 94:67–79

Kankaanpaa H, Vallius H, Sandman O, Niemisto L (1997) Determination of recent sedimentation in the Gulf of Finland using ^{137}Cs. Oceanologia Acta 20(6):823–836

Kuliński K (2010) Obieg węgla w Morzu Bałtyckim (Carbon cycling in the Baltic Sea). Ph.D. thesis. Institute of Oceanology PAS, 134 pp (in Polish)

Kuliński K, Pempkowiak J (2008) Dissolved organic carbon in the southern Baltic sea: quantification of factors affecting its distribution. Estuar Coast Shelf Sci 78:38–44

Kuliński K, Pempkowiak J (2011b) Accumulation, mineralization and burial rates of organic carbon in the Baltic sea sediments. Marine Chemistry submitted

Kuliński K, She J, Pempkowiak J (2011) Short and medium term dynamics of the carbon exchange between the Baltic Sea and the North Sea. Cont Schelf Res 31:1611–1619

Kuliński K, Dzierzbicka-Glowacka L, Maciejewska A, Pempkowiak J (2011) Parameterisation of a zero-dimensional Pelagic detritus model, Gdańsk deep, Baltic sea. Rocznik Ochrona Środowiska 13:187–206

Lima AL, Hubeny JB, Reddy Ch, King JW, Hughen KA, Eglinton T (2004) High-resolution historical records from Pettaquamscutt River basin sediments: 1. ^{210}Pb and varve chronologies record of ^{137}Cs released by the Czernobyl accident. Geochimica and Cosmochimica Acta 69(7):1806–1812

Martin WR, McCorkle DC (1993) Dissolved organic carbon concentrations in marine pore waters determined by high-temperature oxidation. Limnol Oceanogr 38:1464–1479

Müller A (2002) Organic carbon burial rates, and carbon and sulfur relationships in coastal sediments of the southern Baltic sea. Appl Geochem 17:337–352

Nausch G, Nehring D, Nagel K (2008) Nutrient concentrations, trends and their relation to eutrophication. In: Feistel R, Nausch G, Wasmund N (eds) State and evolution of the Baltic sea 1952–2005. Wiley, Hoboken, pp 337–393

Pempkowiak J (1991) Enrichment factors of heavy metals in the Southern Baltic surface sediments dated with ^{210}Pb and ^{137}Cs. Environ Int 17:421–428

Prowe AEF, Thomas H, Pätsch J, Kühn W, Bozec Y, Schiettecatte L-S, Borges AV, de Baar HJW (2009) Mechanisms controlling the air-sea CO_2 flux in the North Sea. Cont Shelf Res 29:1801–1808

Robins JA (1978) Geochemical and geophysical applications of radioactive lead. In: Nriagu JO (ed) The biogeochemistry of lead in the environment. Elsevier, Amsterdam, pp 253–393

Rullkötter J (2006) Organic matter: the driving force for early diagenesis. In: Schulz HD, Zabel M (eds) Marine geochemistry. Springer, Berlin, pp 125–206

Schmidt M, Fennel W, Neumann T, Seifert T (2008) Description of the Baltic sea with numerical models. In: Feistel R, Nausch G, Wasmund N (eds) State and evolution of the Baltic sea 1952–2005. Wiley, Hoboken, pp 583–624

Schulz HD (2006) Quantification of early diagenesis: dissolved constituents in marine pore water. In: Schulz HD, Zabel M (eds) Marine Geochemistry. Springer, Berlin, pp 73–124

Sharp JH (2002) Analytical methods for total DOM pools. In: Hansell DA, Carlson CA (eds) Biogeochemistry of marine dissolved organic matter. Elsevier Science, San Diego, pp 35–58

She J, Berg P, Høyer JL, Larsen J, Nielsen JW (2006) On operational three dimensional hydrodynamic model validation. In: Proceedings of EuroGOOS 4th Conference, pp 519–526

She J, Berg P, Berg J (2007) Bathymetry impacts on water exchange modeling through the Danish Straits. J Marine Syst 65:450–459

Suratman S, Weston K, Jickells T, Fernand L (2009) Spatial and seasonal changes of dissolved and particulate organic C in the North Sea. Hydrobiologia 628:13–25

Szczepańska A, Maciejewska A, Zaborska A, Kuliński K, Pempkowiak J (2011) Distribution and provenience of organic carbon in the Baltic sediments. Geochronometria, accepted

Thomas H, Schneider B (1999) The seasonal cycle of carbon dioxide in Baltic sea surface waters. J Marine Syst 22:53–67

Thomas H, Pempkowiak J, Wulff F, Nagel K (2010) The Baltic sea carbon and nutrient fluxes in continental margins. Springer, Berlin, pp 234–245

Ullman WJ, Aller RC (1982) Diffusion coefficients in nearshore marine sediments. Limnol Oceanogr 27:552–556

Umlauf L, Burchard H, Hutter K (2003) Extending the k-omega turbulence model towards oceanic applications. Ocean Model 5:195–218

Widrowski H, Pempkowiak J (1986) The history of surface sediments in the Southern Baltic. In: Proceedings 15th Conference Baltic Oceanograph. Marine Pollution Laboratory, Copenhagen, pp 656–671

Zaborska A, Carrol J, Papucci C, Pempkowiak J (2007) Intercomparison of alpha and gamma spectrometry techniques used in Pb-210 geochronology. J Environ Radioact 93:38–50

Zaborska A, Carrol J, Papucci C, Torricelli L, Carrol M, Walkusz-Miotk J, Pempkowiak J (2008) Recent sediment accumulation rates for the Western margin of the Barents sea. Deep-Sea Res II 55:2361–2371

Zeldis J, Hicks M, Trustrum N, Orpin A, Nodder S, Probert K, Shankar U, Currie K (2010) New Zealand continental margins. In: Liu K-K, Atkinson L, Quiñones RA, Talaue-McManus L (eds) Carbon and nutrient fluxes in continental margins. Springer, Berlin, pp 273–287

Index

A

Arkona Basin, 85, 88, 90, 109, 111
  carbon return flux from sediments, 88
  emission of CO_2, 85

B

Baltic-C, a BONUS project, 65
Baltic Sea
  Arkona Basin, 85, 88, 90, 109, 111
  as a diverse system, 2, 49
  as sink of CO_2, 86
  atmospheric deposition of carbon, 76
  Baltic Proper, carbon sedimentation, 87
  Bornholm Deep, 68, 114
  Bothnian Bay and Gulf of Bothnia, 2–3, 17, 28–30, 32–33, 35–36, 67, 104
  box model of carbon fluxes, 49–50
  carbon budget, 86
  characteristics, 28
  climate change consequences, 17
  CO_2 balance for Baltic, 77
  CO_2 net emission from, 82
  coastal erosion, 83
  Danish Straits, 28, 31, 34, 52, 55, 88, 109
  deposition and provenience of organic matter, 118
  depositional basins, 95
  doubts regarding the role as source/sink of carbon, 28, 49
  emission/absorption of CO_2, 83
  estuaries, 85
  exchange of carbon with North Sea, 103
  export of carbon to North Sea, 90
  Gdansk Deep, 68, 88, 95, 114
  Gotland Deep, 69, 88, 95, 114
  Gulf of Gdansk, 112
  Gulf of Riga, 82, 114
  inflow of CO_2, 81
  Kattegat, 109
  nitrogen in sediments, 117
  nutrients in seawater, 86
  outflow of CO_2, 81
  point sources of carbon, 36
  river influence on CO_2 emission from, 85
  rivers as carriers of carbon, 63
  sea level rise, 17
  Skagerrak, 109
  Southern Baltic, 85
  summary of carbon gains and losses, 86
  transformation of hydrological regime, 102
  trophic status, 85, 92
Biological pump
  definition, 1
  mechanism, 24–25, 90
  relation to nutrients increase, 93
  significance in carbon cycle, 1, 26, 93, 113
Bornholm Deep, 66, 114
Bothnian Bay, 2–5, 17, 28–30, 32–34, 36, 103, 114
Box model of carbon fluxes, 49–50

C

Carbon
  accumulation in Baltic sediments by regions, 87
  atmospheric deposition to Baltic, 76
  black, 9
  budget in the Baltic, 49–51, 86

K. Kuliński and J. Pempkowiak, *Carbon Cycling in the Baltic Sea*,
Geoplanet: Earth and Planetary Sciences, DOI: 10.1007/978-3-642-19388-0,

Carbon (*Cont.*)
a scheme of carbon in the Baltic, 82
factors determining, 81, 87
uncertainty, 90–91
burial, definition, 114
changes on early diagenesis, 88
changes on sediments incubation, 72
deposition to the Baltic sediments, 66, 87, 112, 114
factors controlling, 87
in selected areas of the Baltic, 67
DIC, 23, 25, 30–33, 51, 58–61, 112
concentrations in pore water, 67, 69, 87
concentrations in sea water, 58–59
export to the North Sea/import to the Baltic, 59
load in river runoff, 64
measurements, methodology, 112, 114
measurements, precision, 115–116
return flux from sediments, 87
seasonal fluctuations of, 58–59
diffusive flux from sediments, 70
dioxide (CO_2), 1–5, 8, 11–13, 15–16, 19–28, 35–36, 90
DOC, 23–24, 26, 30–33, 51, 61–63, 112
concentrations in pore water, 67, 69, 88
concentrations in the Baltic/North Sea, 62
export/import to the Baltic, 62
in the near shore zone, 62
load in river runoff, 63
measurements, methodology, 112, 114
measurements, precision, 115–116
return flux from sediments, 87
seasonal variability in Baltic and North Sea, 61
exchange between Baltic and North Sea, 31, 37, 50–63, 102
exchange between land and sea, 22, 37
fisheries as a carbon loss in Baltic, 36–37
inflow to Baltic with precipitation, 75
inflows from the North Sea, 86
inflows to Baltic with river runoff, 63, 86
inflows to/outflows from, Baltic, 82
influence of climate change, 102
inorganic (IC) and organic (OC), in Baltic, 86, 103
largest source to Baltic, 86
load in rivers
loss from shelf seas sediments, 88
losses from sediments due to mineralization, 74, 88
organic, concentration in sediments, 71
organic, loss rate, 72
PIC, 23–24, 33
PIC, POC as carriers of carbon to sediments, 24, 33–34, 51
POC, 23–26, 30, 33–34
point sources of carbon to Baltic, 36
POM, 114
precipitation as a source of carbon to Baltic, 36
reservoirs in the Earth's crust, 19
return flux from sediments to water column, 67
factors influencing, 88
fraction of total carbon, 88
in various Baltic regions, 88
range of carbon loss from sediments, 88
sources and losses in Baltic, summary, 86
stable isotopes, 19
terrestrial carbon input, 37, 51
TIC and TOC
in river runoff, 63
return flux from sediments, 70
Carbon cycle
as part of matter cycle, 20
exchange between land and sea, 21
global, 1
homeostasis in cc, 21
human influence, 22
in Baltic, 49–50, 86
natural carbon cycling, 1
Carbon dioxide (CO_2), 1–5, 8, 11–13, 15–16, 19–28, 35–36, 90
absorption/emission by shelf seas, 83
as a cause of pH decrease, 103
as a closing term of Baltic carbon budget, 37–50
bilance for Baltic, 77
cause for absorption/emission discrepancy, 83
concentration change consequences, 101
concentration in rainwater, 76
diffusion flux from sediments, 87
emission/absorption by North Sea, 84
emissions from Baltic, 81–83
emissions from estuaries, 84
exchange between sea and atmosphere, 2, 21, 25, 35–50
in rainwater, 76
in shelf seas, data discrepancy, 84
net emission from the Baltic, 81
partial pressure, 2, 14, 23, 49
production, 25
return flux from sediments, 88
river influence on emission, 85
seasonal variations in seawater, 58

summary of emission from shelf seas, 84
Climate
aerosols, the influence on, 9
albedo significance, 10
categorization systems, 5
change
influence on Baltic carbon cycle, 101
influence on extreme events, 102
as cause of river runoff increase in Baltic, 102
temperature increase on carbon cycling in Baltic, 102
change; consequences to carbon budget, 101
change; consequences to the Baltic ecosystem, 18
change; extreme events, 16
change; in the Baltic region, 17
change; society actions against, 16
change; species extinction, 16
definition, 5
extraterrestrial and planetary factors, 6
greenhouse effect, 8, 11, 19
greenhouse gasses influence on c, 8–9, 11, 17
high and low pressure systems influence on c, 8
temperature increase, 8, 13–14, 19
types, 5
Coastline erosion, Baltic, 103

D
Detritus, 23–24, 30
DIC, 23, 25, 30–33, 51, 58–61, 112
see in carbon
Diffusive flux of carbon from sediments, 70
DOC, 23–24, 26, 30–33, 51, 61–63, 112
see in carbon

E
Elemental Analyser Flesh EA 1112 Series, 117
End members method, 111

F
Fossil fuels, 9–10, 12–13, 20, 22

G
Gdańsk Deep, 68, 88, 95, 114
Gotland Deep, 69, 88, 95, 114
Greenhouse gases, 8, 11, 17, 101
Gulf of Bothnia
emissions of CO_2, 82
primary productivity, 85
Gulf of Riga, 82

H
HELCOM, 2, 17–19, 27–30, 33, 36
Henry law of diffusion, 77

I
Incubation of sediments, 71
IPCC, 1, 5, 8–15, 17, 19, 22
Isotopic Ratio Mass Spectrometer IRMS, 117

L
Labile/resistant organic matter
decomposition of, 70
in Baltic sediments, 70, 87
Loss of organic carbon from sediments
on incubation, 72
rate, 74

M
Miami, Univ., 113

N
Nitrogen
accumulation rates in sediments, 95, 117
as important biogenic element, 94
load deposited to sediment, 97
North Sea, 85, 103, 109, 111
emissions of CO_2, 85
exchange of carbon with Baltic, 88, 103
import of carbon from Baltic, 90
Nutrients in Baltic, 101
climate change influence on concentration, 103

O
Oceania, rv, 112
Organic matter/carbon, 1–5, 10–12, 21, 23–26, 28, 30–34, 51–55
accumulation in sediments, 51
allochtonous/autochtonous in the Baltic, 93
burial, 2
deposition and provenience in the Baltic, 93
in aerosols, 10

Organic matter/carbon (*Cont*)
labile in sediments, 70, 94
load in precipitation, 76–77
loss from sediments, 74
importance, 89
loss from sediments due to mineralization, 87, 89
mineralization, 1–2, 24, 26, 28, 30, 33–34, 51–55
of organic matter in Baltic sediments, 87, 116
provenience in sediments, 98, 116
refractory in sediments, 73, 94
terrestrial organic matter, 8, 31–32, 35
stable carbon isotopes, 100
terrestrial; contribution to total organic matter, 100

P

pH of seawater
influence on phytoplankton, 103
shift on climate change, 103
Phosphorus
C/P ratios in sediments, 74
in sediment cores, 74
Photosynthesis, 1, 20–22, 24–25, 30
Phytoplankton, 1, 9, 24, 26–27, 30
Precipitation
carbon in, 75
CO_2 in rainwater, 76
load of organic carbon, 76
organic carbon in precipitation, 77
uncertainty, 90
Primary production, 1–2, 16, 21, 24–26, 28, 33

R

Radiocesium, 95, 116
Radiolead, 95, 116
Rates
organic matter loss from sediments, 74
sedimentation, 87, 95
sedimentation, uncertainty, 91
River runoff to the Baltic, 66

S

Sea level rising, 14
in the Baltic, 18
paleoclimatic reconstructions, 15
thermal expansion, 14
Sea water
concentrations of DIC in, 58
concentrations of DOC in, 62
exchange between Baltic Sea and North Sea, 31, 55–58
pH of, 15–16, 23, 32
sea water acidification, 15, 19
Sediments, Baltic, 1, 20, 24, 26, 33–34, 51
activity concentration of ^{210}Pb, and ^{137}Cs, 95
carbon loss rate from, 88
carbon return flux from sediments, 37, 51–55
concentration of organic carbon in, 71
deposition of organic matter, 93
deposition rate, 87, 95
digenetic processes in, 26, 51–55
incubation for labile matter mineralization, 72, 89
labile organic matter in, 70
linear/mass accumulation rate, 95
N and P concentrations, 74
changes on incubation, 75
molar ratios in sediments, 74
organic carbon accumulation rate, 95
PIC source to, 24
POC as carbon carrier to sediments, 25
provenience of organic matter, 93, 98
refractory organic matter, 73
stable carbon isotopes in organic matter, 100
terrestrial contribution to organic matter, 100
Shelf seas
absorptions/emissions of CO_2, 82–83
as a source of CO_2, 29
cause of CO_2 emissions discrepancy, 83
of Atlantic, 85
of Pacific, CO_2 absorption, 83
the role in carbon cycling quantification, 28
the role of shelf seas in carbon cycling, 2
Stable carbon isotopes, 101
as indicators of organic matter provenience, 102
in sedimentary organic matter, 101

T

Thermo Electron Corporation, 117

U

Uncertainty
Baltic carbon budget, 89–92

carbon concentrations in sediments, 70
carbon deposition to sediments, 91
carbon exchange Baltic/North Sea, 90
carbon in precipitation, 91
carbon load in river runoff, 91
concentrations of DOC/DIC, 90
of net CO_2 exchange
with atmosphere, 92

W
World Meteorological Organization (WMO), 5
World Ocean
circulation system, 9

Z
Zooplankton, 23–24, 26

MIX
Papier aus verantwortungsvollen Quellen
Paper from responsible sources
FSC® C105338

If you have any concerns about our products,
you can contact us on
ProductSafety@springernature.com

In case Publisher is established outside the EU,
the EU authorized representative is:
Springer Nature Customer Service Center GmbH
Europaplatz 3, 69115 Heidelberg, Germany

Printed by Libri Plureos GmbH
in Hamburg, Germany